D1171156

Dynamic models in earth-science instruction

Dynamic models in earth-science instruction

W. H. Yoxall
Professor of Geography and Earth Science,
University of Malawi

Cambridge University Press
Cambridge
London New York New Rochelle
Melbourne Sydney

Published by the Press Syndicate of the University of Cambridge
The Pitt Building, Trumpington Street, Cambridge CB2 1RP
37 East 57th Street, New York, NY 10022, USA
296 Beaconsfield Parade, Middle Park, Melbourne 3206, Australia

First published 1983

Printed in Great Britain at the University Press, Cambridge

Library of Congress catalogue card number: 82-4385

British Library cataloguing in publication data

Yoxall, W. H.
Dynamic models in earth-science instruction.
1. Earth sciences
I. Title
550 QE26.2
ISBN 0 521 24262 2

Contents

Preface

In earth-science undertakings, field study and laboratory analysis and testing are always integrated – the geologist, soil engineer, hydrologist, geomorphologist and civil engineer will all carry samples and data back to the laboratory and examine field concepts by experimental means.

One of the most valuable techniques is to construct dynamic models of features such as estuaries, harbours, dam sites or river segments to test for the impact of natural agents on man-made structures. In a more academic vein, considerable work with models has been done by earth-scientists on the essential character of natural processes – such as grading and meandering in river channels, and beach formation in marine environments.

These techniques are used in the present volume to demonstrate equipment and model controls that will develop natural processes and micro-landforms in the laboratory. For simplicity, generalized landform cases are taken, rather than specific landforms, but aspects of scaling-down from reality are given attention in Chapters 3 and 8.

Models of this kind may be used at a very elementary level and originally it was intended to present ranges of sophistication. Time and space have narrowed the presentation to a level appropriate to upper high school or introductory university courses. The equipment used is mainly the flume and rainfall simulator, which were developed in research situations and have a wide range of application. Up to a point, the equipment gains in accuracy and flexibility as it is designed and constructed with more care, but either piece of apparatus may, if necessary, be reduced to a simple core system.

The rainfall simulator in particular repays attention given to design, and, as it is much less common in laboratories than flumes or stream tables, extensive reference to points of instrumentation taken from the era of basic rainfall simulator development has been included in Chapter 5. It is a trickier instrument than the flume, and its abilities need to be learned more carefully.

The book uses a simple systems approach to focus attention upon the whole and to aid in synthesizing structured explanations. Systems vocabulary is kept to a minimum and concepts are introduced only in those instances where they are most apt.

Equipment has been produced and housed through the years with the help and support of many people. My special thanks are extended to Mr K. J. Sage, Chief Technician at U.C.C., Ghana and the late Professor K. A. J. Nyarko; to Professor Joseph Haughton of Trinity College, Dublin and Professor Paul Williams of the University of Auckland.

At Memorial University, Newfoundland, my colleague Professor Maurice Scarlett, as Head of Department, made possible extensive expansion of our laboratory facilities, and technical assistance was given by the Technical Service section and by Mr J. Everard, department technician; Miss Josephine Ryan and Mrs Glenda Rollings typed the manuscript with considerable fortitude.

I should like especially to acknowledge the many students who have, in various institutions, joined me in experimental work – sometimes as observers, sometimes as guides to groups of visitors, sometimes as demonstrators or research workers, becoming involved with innovation and feeding back ideas and equipment into the teaching side of laboratory use; it is these students who most clearly appreciate the linkage and interdependence of research and demonstration.

Introduction

The use of laboratory equipment in earth-science coursework is becoming commonplace, but, in schools at least, is still sufficiently dependent upon the initiative of individual instructors to be affected fundamentally by the vagaries of budgets and administrative awareness. It doesn't enjoy the 'essential-to-include' image of experimental work in the sister sciences of chemistry, physics and biology. And often, when experimental work is included in earth-science courses, it involves an elementary approach that has little concern with analysis or hope of replication – this is especially true of stream-table work based on commercial equipment.

This book is addressed particularly to the upper grades of high school and to early university courses, and sets out to describe apparatus offering experimental accuracy and flexibility. The apparatus described has been developed and used by the writer over a number of years in various institutions, and is mostly made on the premises.

While no attempt is made to outline actual coursework or syllabuses, examples of experiments are given in the formal 'objective', 'procedure', 'conclusion' format, and in every case an attempt is made to re-assess the conclusions within a systems framework. Since, in general, results for instruction can be obtained using a fair range of hydraulic parameters, specific Froude or Reynolds numbers are rarely given. However, these, together with bed roughness, slope and discharge are discussed where appropriate, but particularly at the beginning of Chapter 8.

It is strongly felt that there is as much value in laboratory work using dynamic models for earth-science instruction as in using long-standardized laboratory apparatus for studies in chemistry, physics and biology. Earth-science experiments, however, fall into the inexact science category, and few experiments can be as clear-cut and theoretically explicit as, say, those in elementary physics. Even so, there is a general body of understanding concerning most processes in earth science and interpretation in this work has attempted to follow non-controversial explanation without constant reference to authoritative sources, since discussion of basic processes is always to hand in school and college texts. The inevitable shades of difference in interpretation and the questions not immediately answerable, form part of the attraction of work in this field.

1 Models

Models of varying kinds are playing an increasingly important part in intellectual life, and this is particularly so in the sciences. It has been said that in the social sciences models are an essential part of any research project,[1] and models in a practical, engineering context are commonplace.

It is usually difficult, if not impossible, to comprehend the complexity of reality unless it is reduced to certain simplified terms which abstract what are believed to be the essentials. This is true of geomorphic and earth-science models, and in this field models have been in use since early times. The simple act of tracing a symbolic stream or mountain barrier in sand is producing a model, and a map is a similar model construct.

In play, models are of considerable value – as with the scenery of a model railway, the model railway itself, the hills upon which toy castles stand, and models of cars or babies in which form, at least, is preserved in scale. Closer to the point are static models of relief form or geologic structure, made of papier mâché, clay or plastic: these are widely described elsewhere, but may form the core of dynamic models.

The present concern is specifically with dynamic models which make use of materials in motion for required effects. This type of model is commonly employed for demonstration or research, and the relationship between these two goals must be examined.

Dynamic models for demonstration have a special attraction – they are concerned with changing conditions, which is an important factor in the holding of attention and the creation of interest.[2] Aspects of this have been noted by others, for instance by such means reality is said to be illuminated,[3] and, if students participate in setting up and manipulating models, the involvement creates a valuable learning situation,[4] and 'Models are of the greatest value in teaching, particularly if the students are involved in their design, construction and operation. . . . The process of making, doing, observing and measuring develops the student's intuitive capacity, and relationships and principles are often perceived long before he is able to understand a complete and regional explanation of what he is observing or creating.'[5]

Perhaps one of the most important aspects of dynamic hardware models is their ability to represent landform features or associated processes not normally existent in a given location. For instance semi-arid landforms will not be available for inspection by students in wet temperate climates, nor may certain types of stream channel. Of special value also, in geomorphic and earth-science contexts, is the ability of a model to represent a total spatial concept which may be appraised more or less by standing in one spot. Very few models are so large that the entire scheme is not visible from a single vantage point. This is of great value for its visual impact and conceptually in its representation of the holistic nature of earth phenomena. The likelihood that many students will eventually undertake fieldwork on parts of the whole (perhaps very small parts) makes it more advisable to instil an appreciation of holism.

The position of dynamic models within the general set of models may be regarded from two positions, the nature of the model type and the use to which it is put.

Any classification involving the nature of models should ideally accommodate all possible models. The prior identification of what a model is, is a matter of great complexity. It has been pointed out that some definitions of the term are so loose that it almost becomes meaningless.[6] This is especially the case where the concern is with theory or hypothesis. With hardware models, no such problems arise. Whatever the definition of model adopted, the visual representation of reality, of a prototype by like form, is understood to fall squarely in the model field. In fact, the image – the scaled reproduction that is visually similar to reality – is the archetype of the model concept.

What is of importance is the placing of hardware models into an appropriate framework or classificatory structure.

A definition of model may include theory or law or hypothesis or structured idea, and may include a role, a relation or an equation. 'It can be a synthesis of data ... it can also include reasoning about the real world by means of translations in space (to give spatial models) or in time (to give historical models).'[7]

From general considerations it is possible to delineate two categories of model with respect to their substance, that is, of what they are made;[8] these are the physical and the theoretical. The physical, which may also be regarded as a hardware model or an experimental model, is the type to which all examples in this book belong. The theoretical model, which is concerned with symbolic representation of reality, or mental concepts, is of no special concern in the present work.

The *nature* or *stuff* of a model assumes particular importance in our terms when the use of the model is also considered. One of the problems met in using dynamic models in instruction lies in explaining the extent to which the model is *true to life*. If the model is to be used at a very elementary, visual-impact level, the actual materials used need not be of concern.

If, however, the model is viewed in an analytical sense (and the occurrence of processes will lead most intelligent observers to think analytically), the nature of materials (or energy form or colour) becomes very important. Models may use the *stuff* of the prototype or else some substitute. If prototype substances are used then the model is *iconic*; if substitute substances are used the model is an *analogue*. It may be noted that some of the models termed *theoretical* above might also be regarded as analogue models, for example, if symbols are used to represent prototype factors.

1.1 The use of the model

Model classification – hardware model classification – is usually concerned with placing models into a research framework (Figure 1.1).[9]

The model for models of Figure 1.1 can be adapted to models for demonstration by using the methodology shown in Figure 1.2, in which the 'observations' are used for conceptual understanding. 'Translation' (T_{3T}) and 'mathematization' (T_{3M}) are methodological steps not necessarily taken. Also, as we are concerned with dynamic models, the methodological step 'experimentation' (T_{4E}) is carried out but observations are not *necessarily* data collection and 'rationalization' may follow rather than 'statistical' (i.e. quantitative) 'interpretation' (T_{5E}). It is at this point of 'observations' that the model used in demonstration may be treated differently from that used in research. There may, however, be research involved in using a model for demonstration, that is, in terms of collecting data for *perception* analysis, or in bringing a model to an optimum simulation condition, in which cases attention has to be paid to quantitative data and the fusion between demonstration and research becomes apparent, and classification complex.

Frequently, the 'conclusions about the real world' are not derived from quantitative observations, and this situation seems to be covered by Chorley's statement,[10] 'It is interesting that many lines of reasoning in geography reach these conclusions (i.e. conclusions about the real world) by a process of more or less direct reasoning (T_D).'

The applicability of this statement may be seen to be unfortunately common in model demonstrations as shown in Figure 1.3.

Such methodology is as open to criticism as 'eyeballing' the prototype.

Models used in demonstration should be approached more analytically by following a procedure such as suggested in Figure 1.2. Measurements may or may not be taken, conclusions will be based on quantitative data or simply rationalization, but in either instance adequate background knowledge will have to be applied, and more attention should be given to holistic concepts than may be the case in research into detail.

Figure 1.1. A model for models. (From Chorley (1964).)

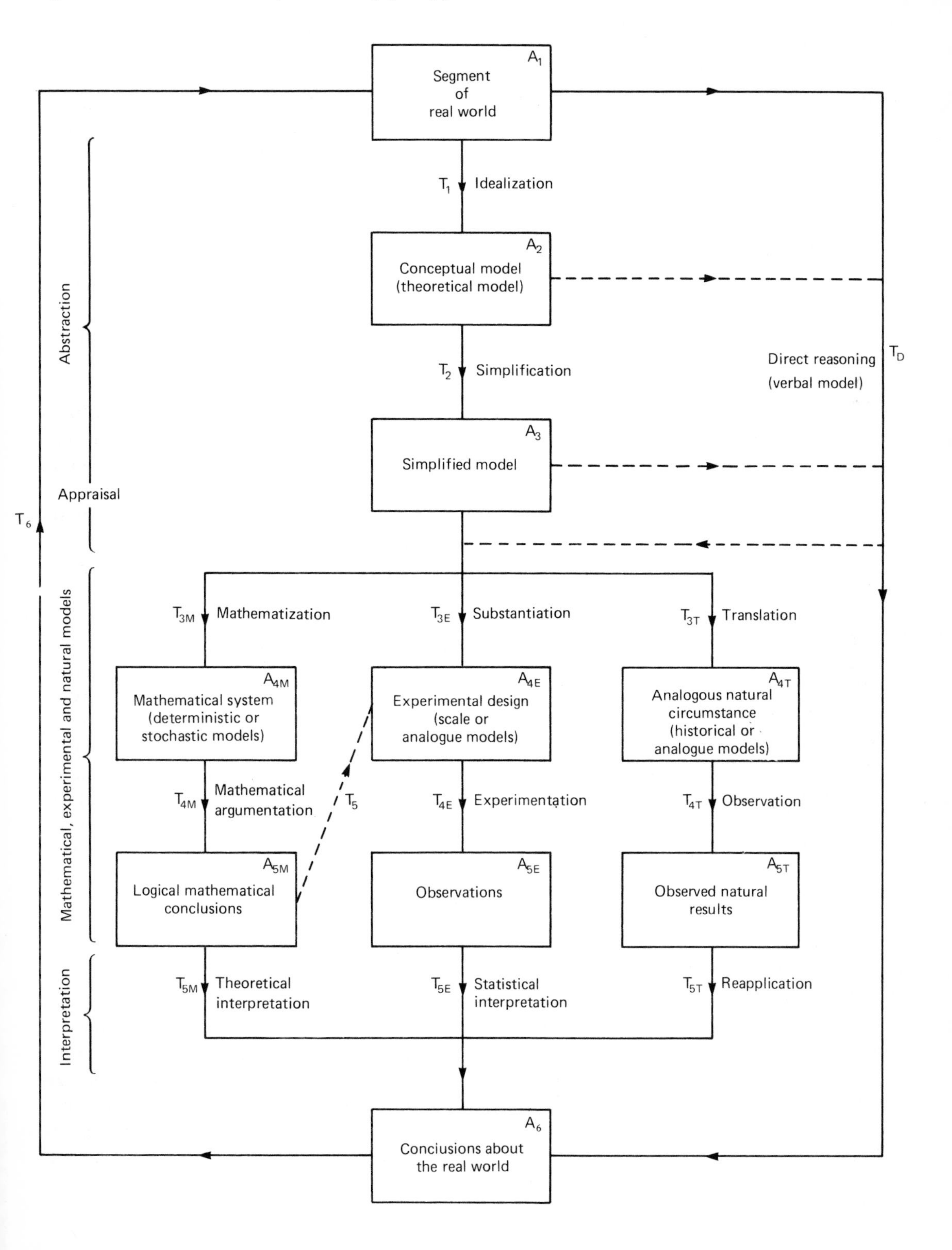

The situation may be summarized as shown in Figure 1.4.

As suggested, the line between demonstration and research is very difficult to draw. It would be a singularly unimaginative piece of research work with a dynamic model that did not reflect qualities of process beyond the instrumented data collection, and in demonstration overall conceptual understanding is greatly enhanced by collection of quantitative data.

Theory of models for demonstration must recognize a point of divergence from research in this data-collection area, where models in research tend to focus upon isolated process parameters, while models for demonstration tend to be used in a broader way, in terms of concepts, identities and interpretation.

The extensive areas of overlap, in which it is impossible to separate demonstration from research categories, include the situation in which a research experiment is used as a demonstration. It frequently happens that a person involved in research is called upon to explain the purpose and types of result to a general audience. For instance, the rainfall simulator eroding soil samples ceases to be an abstraction, an idealization, a simplification and a substantiation directed towards some specific, limited goal, and becomes for an audience a simple process model of rain falling upon a small field patch in which can be seen the effect of rainsplash, overland flow and sheetwash.

Or at a somewhat more sophisticated level, the same experiment may be run in its proper conceptual context for students with appropriate background who are more aware of the processes and parameters involved but who still are concerned with observing a dynamic model of soil erosion

Figure 1.2.

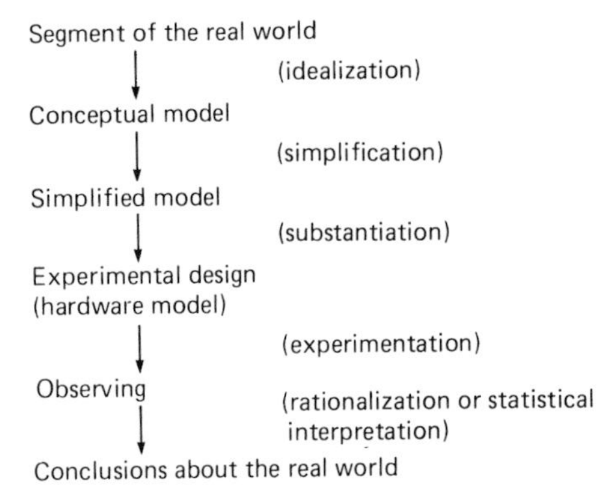

Figure 1.3.

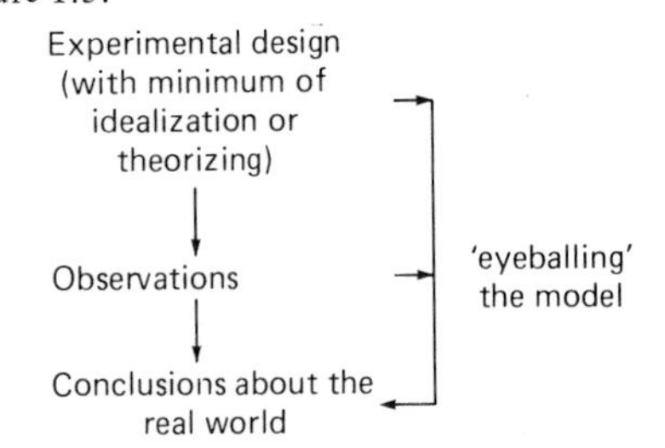

Figure 1.4.

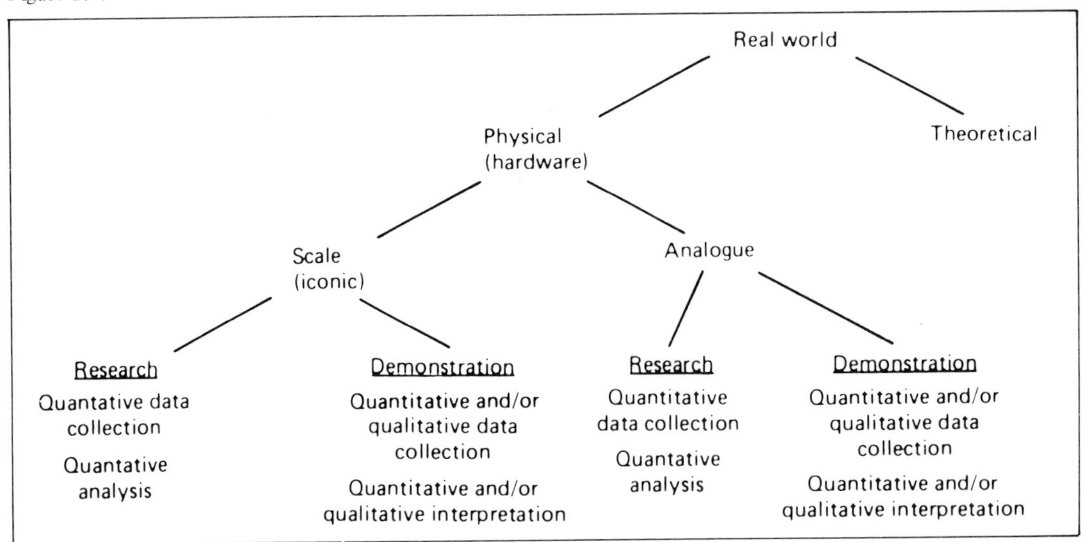

rather than collecting data for research purposes.

While some hydrodynamic models used for research are ideally suited to demonstration - notably those concerned with coastal processes - it is usually necessary to design a model for demonstration with attention to effect. This is often based on adjustment of rates of process.

A situation which must be familiar to all experimental scientists is that in which a research model has to be demonstrated to a lay observer who is manifestly bored after a few minutes - because nothing seems to be happening! Yet if the controls were varied the research worker knows that something could be made to happen that would be of immediate interest. In such cases control changes suited to demonstration rather than research are a matter of degree rather than type.

2 The educational aspects

Models, and especially dynamic models, are capable of demonstrating complex phenomena in a very significant manner. This special educative property has been referred to above for models in general. The impact of a model is so great that it is often regarded as an end in itself - if it is seen, then that is enough.

Properly constructed dynamic models require space, initial funding and considerable instructor-time. The proper use of models should be allowed for in course design, and a given model may be approached in a variety of ways so that it fits comfortably into the overall instructional framework.

Courses in earth science vary from institution to institution, from instructor to instructor and from level to level. There are, however, certain well entrenched approaches into some of which most courses will fit.

If groups are in high school there is a strong possibility that the course will be descriptive and systematic. In such cases it would probably be unwise to use models too analytically. The primary aim would be to show specific form and process phenomena, without looking too closely at the underlying mechanics or the complexity of interaction. There is no doubt, however, that students would be stimulated into asking why or how. They would certainly be curious about cause and effect and would very likely speculate about processes and events using their knowledge of physics and everyday mechanical experience as an hypothesis source. This reaction is common and suggests both a value for the use of dynamic models, and a query regarding courses that are basically descriptive.

In some high school, and most university courses, process-oriented earth-science studies are now standard practice. This calls upon a science background and makes use, by transfer of knowledge, of the concepts of physics and mechanics. It has often been noted that students oriented towards the arts do not approach models with the facility of science students.

When process studies are undertaken the matters of quantity, direction and duration automatically assert themselves, and the desire to check impressions and to test hypotheses becomes imperative. It follows that most effective use of dynamic models in process-oriented courses will involve measurement (quantitative data collection), and the ruler, the balance and the stop-watch should be a corporate part of model study in this context.

A more recent philosophical development for teaching science subjects, including earth-science themes, is holism, using the vehicle of the systems approach.[1,2] This is conceptually satisfying but in practice is difficult to apply in introductory courses. There is a time-conflict between developing the systems construct and imparting sufficient systematic, substantive material. What may occur is that individual themes may be taught as open systems, and integrated at first in pairs and eventually in groups.

Within the limited opportunities offered by thematic open system examples, the dynamic model assumes the additional role of an exceptionally effective vehicle for systems thinking. Any given experiment demonstrating a process may be regarded as a process-response system, derived from morphological and cascading subsystems. There is clearly visible evidence of inputs, work done, outputs and of positive - and negative - feedback loops, dynamic equilibrium (briefly) and static equilibrium more frequently. Also, the concepts of entropy and equifinality may be approached through model use. It is recommended that the powerful concept of systems analysis should be incorporated into model use whenever possible.

Having determined the correct degree of model sophistication for model use appropriate to the nature and level of course being taught, it becomes necessary to integrate models into course work. This means (*a*) planning ahead at the syllabus formulating and organization stages in order that suitable points are identified for demonstration, and (*b*) allocating time for model use. This will be regarded as practical work, laboratory work or perhaps as a supplement to field work. The laboratory period in earth-science courses may exist as a *paper-laboratory* period, and it is in connection with this that model studies might best be used. A two-hour period is probably the most suitable unless students are to set up models themselves, when a longer practical period will be required.

Organization of model study periods is problematical. Often there are fairly large groups of students, and if there are more than twelve in a group it may be best to interlock the model study with a paper-laboratory exercise and to divide the group into subgroups which change about. In any event, models should be large enough for groups of twelve to stand around. If apparatus is small, then there is an advantage in having several units simultaneously at work.

(*c*) The model should never be used as simply something to look at. Students always prefer to have their attention drawn to significant aspects of the model, and are much more attentive and observant when formally involved with exercises. At the simplest descriptive level, exercises may be drawn up in which specific phenomena covered in recent lectures (lessons) are to be identified and possibly mapped, and value judgements can be called for. This may be simplified to the extent of making up multiple-choice questionnaires.

At the more sophisticated levels, questions should be more analytical and call upon a more knowledgeable appraisal or interpretation of process. Although certain experiments may be run for observation and inductive speculation only, it will be more customary for students to measure slopes, velocities, masses transported or deposited over periods of time and to collect data analytically, to calculate and to produce a conceptual idea of the total process or system involved.

Again, this will be in conjunction with lecture material, and should supplement paper exercises to do with maps and aerial photographs.

2.1 Student response

Students tend to be quite specific in their reaction to models of gross landforms. They will accept without significant question, or will be highly sceptical. Either reaction is undesirable, and is based upon a lack of understanding that must be modified. The vital impact of hardware or visual-aid material has an emotional as well as an intellectual component, and too strong a reaction in either acceptance or rejection is probably emotional. Once the student has had to come to terms with the model as a model of reality, with its limitations and advantages over the written word, he will become more moderate - more intellectually appreciative or rationally critical. This particular lesson of attitude never has to be learned twice.

It is necessary for the instructor to know why the particular model in use was constructed just that way - he should know the ways in which it models reality of form and produces relevant processes. He should know what compromises in scale or materials had to be made, and he should be very certain of those aspects of the model that may be regarded as valid and those that may not. Invalidity is introduced for a variety of reasons, usually to suit constraints of time, economics or space. This is true of all dynamic models. If the model is for research purposes care is properly taken to include factors important to the specific objective of the research. In models for demonstration, care should be taken to preserve those attributes of particular value in teaching a concept. A balance has to be struck between possibilities, and the instructor should be aware of what they are. The student has no innate objection to analogy, but has a strong objection to analogy passed off as reality. This is to say that the model must be recognized as a model, and in this context is of more value than it would otherwise be.

There are, of course, some models that are parts of 'unscaled reality'.[3] These are natural out-of-door features that may be studied for what they are, or they may be taken as representative of a set, some members of which are of larger size. These require careful orientation of thought and are discussed in Chapter 12.

From the teaching point of view, dynamic models in earth science should be regarded in much the same way that the inclined plane or electroscope is regarded in physics. In the case of earth science, processes are more complex than in physics, and the consequent model form is not restricted to traditional or mass-produced style. In either case use of apparatus or model may be simplistic or sophisticated, as required, and in each case the object is to familiarize students with a scientific concept.

In this context, the matter of perception has to be considered. The learning process, when instructional aids are in use, is based upon the ability of the aid to communicate an idea or concept. Attention leads to comprehension, and with instructional care, this should lead to acceptance.[4]

It has been assumed that the process of instructing the student by means of a model is automatic. However, comprehension varies considerably from student to student, and this is dependent upon the perception of what is occurring. In experiments with a flume, groups of first-year geomorphology students have been found to perceive processes and landforms in a wide variety of ways. Given choices of answer to immediate post-demonstration perception of stream-channel formation, only 52% correctly adjudged or identified channel sinuosity, width/length ratio of segment, width/depth ratio, bed roughness, velocity marker behaviour, laminar flow, long profile slope and channel side slope. If prior instruction is given on what to look for, perception validity increases.

If the model is to be used in a purely descriptive way, attention has to be drawn to the appropriate attributes of the model - although there is a danger that the student will understand the signal while not necessarily observing the attribute.

If the model is to be used for formal study, the exercise should direct students to the recognition and measurement of all relevant attributes; that is to say, the exercise should instruct the student to identify and measure, rather than tell him *what is*. This also represents a more involved procedure than is the case in using apparatus or models in physics. In the latter, the name and function of the apparatus is known, and the observer will probably be called upon to measure the phenomenon from suitably incorporated calibrations. In the earth-science model, decisions have to be made about the type of measurements to be taken, the localities in the model at which they can be taken, the best way to make them (i.e. what instrument or device) and the interpretation of exactly what part of a phenomenon should be included in the measure. For instance, if the phenomenon involves deltaic material, measures will be linear or gravimetric; they will be made at the mouth of the stream, they will involve some kind of ruler and balance, and the identification of deltaic material, exclusive of river bed or cliff slump, must be made.

The differences are evidently related to the properties of an exact and an inexact science, and exercises (or the instructor) should make it quite clear what must be looked for and guidance must be given in interpretation, isolation, measurement and further interpretation, or inductive reasoning.

Consideration must also be given to student feedback. Students are frequently very constructive in terms of ideas for modifying apparatus and improving its use. They come at the theme freshly, and can sec quite clearly where poor, weak or moderate aspects exist.

Note should always be made of the type of problems they have in carrying out the work, and of the ways in which they make mistakes in writing up their results.

2.2 Instructor's response

While a model is being used for demonstration, the presenter is an instructor. There is a certain aura about dynamic hardware models which tends to push them conceptually into the research sphere. This is especially true of models used in university teaching, where research orientation tends to bend the objective of laboratory instruction into one of research. This is quite unnecessary, and the instructor should avoid *selling* a model on the basis of its research potential.

It has been emphasized above that the line between demonstration and research in model use is probably non-existent. Students who are studying a model for their own learning purposes, or the instructor who demonstrates, may well be doing nothing different from another who uses data from the model for research purposes, and nothing is to be gained from taking the attitude that teaching is good but research is better. They are concerned with the same substantive material; one tends to be conceptually broader than the other. and in model use especially it is not a very moot point as to which is intellectually more rigorous or worthwhile, an hour of good teaching or an hour of good research.

It is necessary, however, that when a model is used in an instructional context, the instructor should know its mechanical potential, its validity and weaknesses, and this information should be ready beforehand, every aspect of model use being carefully worked out on a formal, methodological basis.

3 Validity of demonstration models

The dynamic model in demonstration is required to conform to two sets of criteria: one is the scalar similitude attribute which, having presupposed that the model is reproducing processes characteristic of the prototype, is concerned with distortion of prototype parameters. This is significant because it frequently happens that prototype parameters have to be distorted in the model in order to preserve certain process characteristics. The other set of criteria is concerned with visual impact, and in adjusting to this the demonstration model may lose some of its process truth. It is in this balance of objectives that demonstration models may become extremely weakened, and perhaps it is in this context that some feeling against demonstration models may arise.

It is rare for a model used in research to be accused of not producing the processes it sets out to produce, while the model used in demonstration, whose visual impact is strong, may be instinctively regarded as suspect. This may be a justifiable reaction; but if objectives are kept clearly in mind, and the distortion of process truth is understood and explained, suspicion will be inappropriate and allayed.

The requirement of achieving visual impact needs examination. Generally the demonstration model must produce visible results within a few minutes to an hour. This is the common case, although in certain situations it is possible to set up a demonstration model that will be observed over longer time intervals.

The research model, on the other hand, although frequently designed to foreshorten the time factor, is able to operate over periods of time which may produce no readily perceived change in conditions - measurements may be of micro-magnitude - since it is most likely that such process rates are best for the phenomena under investigation. In order to maintain validity, research models make use of natural materials wherever possible, so that the model is likely to be iconic.

In demonstration models, it is very likely that there will be a substitute of materials, for instance unconsolidated sand may be used for solid bedrock, and the resulting landforms that develop are thought of as existing in solid bedrock. It is probably true to say that, when analogue hardware models are used in formal research, they overlap demonstration analogue hardware models. On the other hand, not all demonstration models are analogue, so that in this sense of the *stuff* of models there is also complexity in classification and intrinsic validity.

The effect of materials and scales used in transposing real-world or prototype situations into model form requires particular attention.

Scalar attributes are subdivided into the three groups of geometric similitude, kinematic similitude and dynamic similitude.[1]

(i) Geometric similitude is concerned with linear parameters ($L =$ length) which affect the size or scale or form of the model. Ideally, all ratios between L_m (lengths of model) and L_p (lengths of prototype) should be constant so that

$$\frac{L_m}{L_p} = K.$$

In constructing a model the ratio L_m/L_p is not usually constant, because of the need to produce a certain process characteristic. For example, a common problem in demonstration models is the depth of water in a river channel which, if scaled down in the same ratio as channel width may not give a required flow characteristic - that is to say, for instance, flow may be laminar instead of turbulent. Also, it is very common in demonstration models to exaggerate the vertical scale of landforms to give better relief impact. In fact, in demonstration models the need to make them *look* like the prototype may be the main determinant of geometric ratios - fortunately it being sometimes possible to work within broad, non-critical dynamic limits.

Distortion of form is in such cases not usually regarded as of primary importance, but this is not an acceptable attitude. As far as possible, distortion should be avoided or carefully balanced to preserve process truth. If in a given instance the choice is made to produce a good visual effect with consequent modified process, it will be important to be aware of process change and to indicate this to observers where necessary.

(ii) Kinematic similitude occurs when paths followed by moving particles and the ratios of rates of movement of homologous particles (as between model and prototype) are equal.

When the paths followed by particles are considered, it is evident that, other things being equal, the particles will be guided by the shape of the model (e.g. the landforms). This will clearly depend upon geometric similitude, and if there is distortion in any of the three Cartesian co-ordinates it will be transmitted to particle trajectory. In practice, this aspect of kinematic similitude is well enough preserved. Exaggeration of the vertical scale will produce steeper slopes and particles will tend to move downslope along a different trajectory from that in the prototype, but still they will move downslope and the difference is unlikely to have any special significance.

Distortion of L_m in the horizontal plane is less likely, and particles will tend to move in the same trajectories as in the prototype. If L_m is distorted in the horizontal, then a changed trajectory in plan would be expected. Although these considerations are important, demonstration models are frequently not scale models of a particular landform - they are generalized - and if this is so may be considered to be free from geometric distortion, and the paths of homologous particles may also be considered as truthful.

With respect to velocity, similitude is preserved when the ratio between V_m (velocity of model) and V_p (velocity of prototype) is constant so that

$$\frac{V_m}{V_p} = K$$

for any homologous moving particles. The preservation of velocity similitude in scaled-down dynamic models is frequently difficult or impossible, even if geometric similitude is completely preserved.

Since kinematic similitude is concerned with velocity, it is also concerned with acceleration, mass and time. In a scale model for demonstration, the erosional and depositional processes are speeded up, but the terminal velocity of falling debris, the velocity of flowing water, the celerity of waves, will be different from the prototype homologue. This is due to the fact that distances are smaller, masses are smaller, but acceleration for example, due to gravity, is constant (g) and operates through a shorter time period.

(iii) Dynamic similitude occurs when the ratio of forces in the model and prototype is constant. There are several relevant forces, such as surface tension, gravity and viscosity[2] and they may be expressed as ΣF. Hence dynamic similitude is preserved when

$$\frac{\Sigma F_m}{\Sigma F_p} = K.$$

Since the acceleration due to gravity is constant, but surface tension and viscosity are dependent upon temperature, and acceleration is a function of mass or density, there is considerable opportunity for the ratio $\Sigma F_m/\Sigma F_p$ to vary. This variation will depend not only upon geometric and kinematic similitude, but also upon such controls as water and room temperature, and the physical properties of the materials in use.

3.1 Discussion of similitude in practice

3.1.1 Geometric parameters. These are length, width and depth (height). In small models there may be little important effect due to distortion unless the model is of a specific earth-surface landform. If it is of a specific landform, height will almost certainly have to be increased out of proportion, and possibly stream-channel parameters, in order to carry a discharge that will have some significant effect.

The meaning of this will be seen if an actual case be taken. Kneehills Creek (Alberta)[3] flows through a stream-cut valley 91.4 m deep, 644 m wide at the top, and 161 m wide at the bottom. The stream near Drumheller is 7.0 m wide and 0.3 m deep with a velocity of 0.03 m s^{-1}.

If a square (plan view) model were built to accommodate the total valley width of 644 m, so that it was 0.914 m square, the scale would be 1:704, and this could also be the vertical scale.

The stream would be (7/704) = 0.01 m wide and 0.000 43 m deep. With a velocity of 0.03 m s^{-1} in the prototype, velocity in the model would be (0.03/704) = 0.000 043 m s^{-1}.

If the kinematic viscosity of water (μ) is taken as 0.0093 $cm^2 s^{-1}$ then the Reynolds number (R) is 0.0043 × 0.043/0.0093, giving $R = 0.02$, which is the highly viscous flow of what amounts to a very thin film of water. It would be impractical to expect to produce this with accuracy and, if it were produced, it would not erode a bed of fine sand.

With a depth of 0.043 cm the water film is not much thicker than the mean diameter of fine to medium grains of sand (0.005–0.5 mm). It is clear that in such a case, the stream parameters will have to be increased in order to produce a dynamic effect. This is virtually saying that if a strong visual effect is required it is not possible properly to represent with a small-scale model a stream-valley prototype such as the Kneehills case.

This provides a limiting constraint to perfect modelling for demonstration of precisely the kind that limits research models.

3.1.2 Kinematic parameters. These are velocity and direction.[4] In most instances a model will not be a specific landform but will be, rather, a general case. The movement of particles will normally be of water or sediment. In hydromorphic models water moves down surface slopes, through the body of landmasses to seepage points, and along river channels. It also moves as currents in sea. In each case the direction of water movement is analogous to that in the prototype. If the model uses simulated rainfall, the water will flow overland in trajectories that replicate natural ones; groundwater observes the laws of hydraulic flow, moving in a laminar mode from points of higher to points of lower hydrostatic pressure (or head), and such trajectories are influenced by land bulk and surface topography and a flow-net can be drawn for this movement with flowlines closely resembling those that would occur in a prototype of similar subsurface materials and structure. In stream channels time-averaged flow directions are as in natural channels, with flowlines predictably positioned on bends and across the section, so that there is a proper relationship between flow trajectories and channel geometry, including such complicating forms as point bars and slump-dams.

With coastal models there is a general correspondence of prototype and model trajectories with flowlines influenced by bays and headlands, deeps and shoals, and the trajectory of wave systems also replicates natural trajectories, with refraction about headlands and into the shoaling bays. Also, the mean direction of waves is determined by their transmission from point or line sources.

It is therefore possible to read natural trajectories into model conditions, providing the model is of a generalized case. If a model is of a specific case, and allowance has been made for distortion of linear parameters (i.e. depth or height) the trajectories of water particles can usually still be related to those in the prototype – a possibility exploited in civil engineering.

3.1.3 Trajectory of material other than water. As the modelling material will usually be unconsolidated sand, there will be a preponderance of undercutting and slumping. The slower, lower vertical angle trajectory of movements associated with creep, solifluction, hillwash etc., will not normally be evident, unless models are especially designed to include them.

Wherever the free-surface water/sand interface occurs, sand will wash out. Sand grains will readily slide out of the interface zone at groundwater seepage points, stream edges and sea edge, where lubrication is greatest, and the unstable equilibrium condition of steep slopes is most prone to adjustment. As a consequence of sapping, slumping occurs. The trajectory of interface particles is similar to that in certain natural conditions such as scree slopes or foreset beds. The trajectory of grains in slump processes resembles prototype slumping or, sometimes, rock fall. There is no known simple way of surmounting the problem of limited trajectory range in mass-wasting processes in models for demonstration. Preparation of a model to demonstrate slower mass movement is time consuming, but may for certain courses be worth attempting. When land surface slopes in the model are gentle enough to permit what may be related to sheetwash, as in rainfall simulator work, the geomorphic features are washed out and, although trajectories of particles over surfaces in dynamic equilibrium occur as they do in prototypes, they are not associated with forms of visual merit and here a specifically demonstration-model limitation is met.

As stated, trajectories of water particles in model rivers can be related to trajectories of water particles in natural streams, and the trajectory of sand grains will be similarly truthful. If the model makes use of unconsolidated sand, the stream channels will be rectangular in section (a notable limiting condition) with undercut and caved-in banks. Within this flat-bottomed section, particles will move in accordance with differentiated velocity lines and eddy or turbulence zones, so that sorting occurs and individual grains can be seen to be differentially influenced by water currents, to be deposited or shifted in trajectories similar to those of natural streams.

The same may be said of particles moving upon a coastal model. Trajectories down an offshore slope or delta, and along a beach, are all similar to those in natural situations. There are, however, secondary effects due to capillarity and to-and-fro movements in micro-turbulence which, although they occur in natural conditions, are not those trajectories of which we think in connection with, say, longshore drift. The property of velocity, which is a kinematic factor, will usually be distorted in models - frequently in models for research and certainly in models for demonstration of processes. In the one case there is the need to reduce the time span of natural processes for economic reasons and in the other case there is the need to reduce the time span for visual-impact reasons. These constraints require the use of readily eroded materials in laboratory models. They also require that sufficient water be used, flowing or moving at sufficient rate, to erode and transport within the required abbreviated time span.

Thus, in demonstration models, it will almost always be the case that higher flow velocity and wave frequency are used than would normally occur in the generalized prototype. This limitation on the use of lower velocity appears to have one major repercussion - once a high enough velocity is attained to initiate the erosion processes, these occur quickly, and the model will run its course with more rapidity than is always desirable. Loose sedimentary material very easily adjusts to an equilibrium condition. Since the motive force for hydromorphic models is in the motion of water, rapid incision, lateral cutting and undercutting are experienced. Rate of recession of interfaces is high; the actual movementof sand grains outside the water bodies will be simply governed by the gravity factor and will, because of the small vertical distances involved, be slow in real terms but in scaled-down terms very rapid. To return to the case of Kneehills Creek, if a stream were designed to represent it and eroded until 91.4 m bluffs occurred, they would be 0.13 m high in the model (using the scale of 1:704). A grain of sand of 1 mm diameter falling down the model bluff would have a mass of

$$\underbrace{\tfrac{4}{3}\pi r^3}_{\text{volume}} \times \underbrace{2.7/1000}_{\text{density}} \text{ g}$$

$$= \frac{5.2 \times 2.7}{1000} = \frac{14.04}{1000} = 0.014 \text{ g},$$

and a velocity of

$$V^2 = 2gh \quad \text{or} \quad V = \sqrt{2gh},$$

where $g = 980 \text{ cm s}^{-2}$ and $h = 13.0$ cm.

Therefore

$$V = \sqrt{2 \times 980 \times 13.0} = 160 \text{ cm s}^{-1}.$$

A sand grain of 0.014 g falling off a 91.44 m cliff will have the velocity (ignoring air-resistance)

$$V = \sqrt{2gh} = \sqrt{2 \times 980 \times 91.44 \times 100}$$
$$= 4233 \text{ cm s}^{-1}.$$

Hence the difference between V_m and V_p is given by the ratio 160:4233 or 1:26.5.

Since mass does not affect velocity, free-fall velocities in the model will be much greater than the scale difference would require.

If the terminal velocity in the model particle were in kinematic similitude it would be $4233/704 = 6.01 \text{ cm s}^{-1}$.

Although the various water velocities are better in models when faster, for process and visual reasons the opposite applies to velocities of mass-wasting particles in which a slower movement would be visually more satisfying. The use of time-lapse photography in conjunction with models is designed to overcome this problem, but such a technique should not be necessary and should usually be regarded as an inferior substitute for live experimentation.

3.1.4 Dynamic parameters. With respect to forces operating in dynamic models, it has already been stated that the gravity factor will operate in a manner independent of scale. The result of this will be to increase the rate of processes relative to the scale of the model. For most models constructed for demonstration, the designer has some idea in mind of what the geometric magnitude of the prototype landform would be. This should be formalized when using the model, so that students are able to envisage accurately the real-world landform, and so that proper explanations and interpretations of scale and distortion are possible.

The limitations imposed by high-velocity occurrences due to the constant gravity factor, have to be accepted in some degree. They give individual processes extremely short life, and the overall life of the model may be inconveniently foreshortened. This concerns the rate of approach to equilibrium conditions with given model controls. The forces acting upon unconsolidated sand, to undercut, wash out and transport are much higher than the geometric model scale. Thus, although in terms of hydraulics the flow characteristics may be maintained by preserving the Froude number* of the prototype, this may involve introducing fixed-bed characteristics, and overall development of the model may be inhibited or prevented.

Generally then, flow characteristics and associated forces will exceed those most desirable. The various erosional and depositional processes will be speeded up, and in a short time the model will reach a state of stable, static equilibrium, and it often happens that this state is relatively featureless. One of the most effective ways of preserving dynamic equilibrium (or instability) and thus of preserving significant landform features, is to slowly change the base-level.

In geomorphic processes, the development of landforms is strongly associated with base-level. Streams cut down to it; water tables adjust to it; interfluves recede with respect to it, and coastal erosion progresses above it. The life of the various forms that develop at the base-level/air interface may be prolonged by dropping base-level by lowering sea-level so that forces retain a significant vertical component and are not permitted to become dominantly lateral.

The usual technique for increasing energy is to hinge a flume so that the model may be tilted, lifting the landmass above a base-level. This is a relatively crude method since it is difficult to control or quantify, and introduces a rotational dimension to the model.

Models using the dropping sea-level technique are described later.

The roles of viscosity and surface tension are more notable in a small model than in a large natural feature. Flow in models is very often viscous, when in natural landforms it would be turbulent. Surface tension affects micro-features in free-surface water, so that small islands and bars are modified at the water/land interface if bed

* See Chapter 8 for discussion of flow modes.

material is fine enough; a small surface-tension terrace forms which has no prototype macro-equivalent. Also, along the sides of streams, where slumping has produced sheer cliffs, undercutting is at first a surface-tension plucking phenomenon, before it is enhanced by the power of running water, and if low-density sand, such as coal or perspex is used, these surface-tension effects are commoner and more pronounced.

Surface-tension effects also occur if floating markers are used for velocity measurements, so that dyes are more effective tracers.

The pressure factor may not influence most demonstration models, unless pressure as such is important - as in artesian or spring-issue situations, where seepage may occur but actual free flow is inhibited due to scalar imbalance between pressure head and flow friction. Often, the head is only just capable of overcoming friction, and issue may be due to the process of capillarity. Hence, the scale of an artesian aquifer which might in the prototype produce a geyser or spectacular flow, may produce nothing better than seepage in the model. In cases like this there is little that can be done to enhance the visual impact without reducing process integrity.* Given that process integrity is not critical, pressure can be increased or friction decreased by a variety of methods.

Acceleration has been discussed as a factor of velocity in kinematic similitude, and, although it rightly belongs amongst dynamic factors, its effect in models is primarily in velocity terms.

3.2 Summary of validity

The question 'is it valid?' is seen to be too simplistic for a demonstration model. Validity must be assessed in terms of formally defined concepts of similitude in addition to those of the more general concepts of visual truth and process representation.

Taking visual truth first, this is dependent upon geometric similitude, and if the ratios are kept constant the model will truthfully represent the form of the prototype.

* However, see Chapter 7, Experiment 9.

As perception plays a part in visual impact, and mankind is very aware of landform relief, visual impact in models may be increased by exaggeration of the vertical scales. This is probably the main invalidity of models in landform situations; there does not seem to be any practical or theoretical reason why this parameter should not be emphasized, but if such a reason is part of a model's design then it is possible to depict real-world landforms in which high relief occurs. Since, however, a demonstration model will normally represent a landform type rather than a specific case, this would be an unusual situation.

In order to satisfy time constraints, including the rapid development of landform with heightened process, several deviations from exact replication and scale integrity may be designed into the model. If a type of unconsolidated sand is used to mould initial landforms that are typically indurated rock, the model becomes an analogue - it being seen that sufficient property differences occur between these two bedrocks to cause them to be considered as different materials. When the prototype is sand, or other clastic material, then the model is iconic and validity is preserved. This may be a matter more of theoretical than practical concern, since processes in either case may be truly representative of reality.

When slopes and discharge or precipitation are increased beyond the scale ratio of the horizontal geometric properties of the model, this distortion should be noted. The resulting processes may be fewer in number than in the prototype, but those that occur are reasonably true. For instance, seepage, undercutting, slumping, rockfall, saltation and deposition, occur in the model as they would in nature.

The variety of forms produced by combination of process and initial landform, together with such controls as are used, is infinite. Part of model validity depends upon keeping the processes and form within certain limits.

Some processes such as capillarity and surface tension, that attain eminence in the model but are not important in natural macro-scale landforms, must be identified and if necessary pointed out to the observer. These, fortunately, do not usually affect visual impact, and only produce moderate *form noise*, i.e. forms that assume too much importance at the model scale.

An element of unreality is introduced by process rates and forces being out of magnitude with the other scalar properties of the model. The very fact that what looks like a fair representation of a river channel lying in a valley before the model is motivated, being suddenly subjected to downcutting, whiplashing and insinuating at a very rapid rate is disconcerting, and although the processes and forms developed are exactly what one wanted, and have to be quickly produced to be observed, the fact remains that the dynamics are *too* rapid. The impression of unreality is so strong that it may well detract from the many valid attributes of the model.

Control over rate is very difficult to assume, but control by base-level adjustment is an effective technique, since it permits the operator to govern form development in the vertical plane and, consequently, in the horizontal.

In final summary, dynamic models may be regarded as valid representations of reality in terms of form and many processes. Some processes that would occur in nature may not occur in a given model, thus in holistic terms the model is not totally true, and in terms of rate it may be invalid.

4 General equipment and controls

Much of the experimental work carried out in demonstrations will make use of the flume, rainfall simulator and wind tunnel. These pieces of equipment may be sophisticated and expensive, but in general it is advisable to employ apparatus that may be produced at home. Universities and colleges usually have technical service departments that are capable of constructing wood and metal equipment, but it is not uncommon for instructors or departmental technicians to develop apparatus, design pump systems and construct plumb and wire equipment.

The auxillary apparatus that may be needed, if it is beyond the scope of local manufacture, can usually be bought from ordinary laboratory suppliers at competitive prices.

The justification for avoiding over-sophistication is not only economic, but also conceptual: there is an inherent beauty or appeal in stylish simplicity, as frequently recognized in research,[1] and which is of special value in instructional models where the focus must be on concept, not hardware.

4.1 The bed material

Although fixed-bed or fixed-form models may be used in dynamic system demonstration, they have limited operational utility, and discussion here is restricted to mobile-bed models where landforms may change under the action of water.

In typical demonstration laboratory situations the quantity of water moving through a system must be adequate to modify landforms but not enough to wash the model out too quickly. A factor of importance in achieving these ends is the bed material upon which the water is to work. It should be capable of dislodgement by water, to produce sharp or strong relief features, but should not be too readily moved or too resistant.

These requirements are met by sands with good cohesive properties. A fair range of small grain sizes may be used, but good results are obtained with material in the 0.21 mm bracket. If it is too fine (say <0.125 mm) then side effects such as a low coefficient of permeability and poor seepage may occur, to retard groundwater movement, and if it is too large (say 0.5 mm) then it will not readily be dislodged and transported, so that fine surface features will not develop.[2]

Angular grains will be able to pack and cohere better than rounded, and the presence of particles of different colour in the general matrix greatly aids close scrutiny of trajectories and velocity.

The density of sand employed is as important as grain size. There has been a wide variety of substances used in dynamic models, with the object of scaling down density of sedimentary material to some measure commensurate with other scalar ratios. Granular hardwood particles, boiled to expel air, provide a very low-density grain;[3] granulated perspex has the properties of low density and durability;[4] pumice has also been used to good effect,[5] being a satisfying material to work with. If variations in effect with density are to be demonstrated, the range of sands, coal (s.g. 1.34), silica (s.g. 2.64) and limestone (s.g. 2.72) are convenient. Anthracite is the best type of coal to use and limestone is available from appropriate quarrying companies. A fine silica sand from a beach is probably the most accessible substance; it is fairly uniform in size, clean and relatively cheap.

These bed materials are not readily acquired, since commercial outlets are limited and the substances expensive. It may be best to find a bulk source and to crush and sieve it oneself.

Once the sand is obtained and ready for use in the laboratory it should be regarded as a special substance and neither adulterated with different materials nor wasted by casual spillage.

If a suitable sand is not available and it is necessary to produce it, a material of the correct density is selected and fed, firstly into jaw-crushers (Figure 4.1) and then through a grinding device such as a disc grinder (Figure 4.2) or a rod or ball mill (Figure 4.3), until most of it will pass through the appropriate sieve. This treatment is restricted to brittle, mineralized substances, including coal.

Figure 4.1. Jaw-crushers. (From E. H. Sargent & Co. Scientific Instruments catalogue (1967).)

Figure 4.2. Disc grinder. (From E. H. Sargent & Co. Scientific Instruments catalogue (1967).)

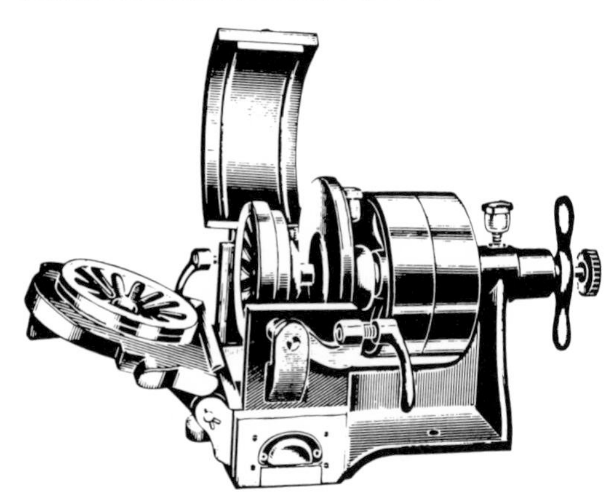

Quantities used in demonstration models are usually substantial, and grinding can be a lengthy and arduous task. If a department in the teaching institution does not own suitable equipment, in say the geology or soil mechanics sectors, it is possible that commercial organizations such as quarrying or concrete manufacturing companies will allow use of their equipment.

Before it can be used in the laboratory, sand must be properly sieved and cleaned to remove fines and organic material which will form a scum on the water surface in models. This involves wet-sieving, which is also a fairly large job, but once done does not have to be repeated.

Wet-sieving requires a plentiful supply of water and a sink unit that incorporates a sediment trap. If the unit is an ordinary laboratory sink, the use of a receiving tank for waste water is strongly recommended. This will have to be syphoned or pumped into the standard sink, care being taken to leave behind sediment and dirt. Low-density substances such as coal are especially troublesome in

Figure 4.3. Ball mill. (From E. H. Sargent & Co. Scientific Instruments catalogue (1967).)

wet-sieving and extra care should be taken to trap fines, which can travel considerable distances in pipes before accumulating to form a blockage.

Standard nested sieves are of no use for wet-sieving large quantities of material and a larger tray may be constructed from sieve mesh obtained from a laboratory supplier (Figure 4.4). If sand is required to fall between two limits, such as 0.35 mm and 0.125 mm, two sieves will have to be used, but as they will be bulky it is convenient to use them in turn.

The water should be fed over the sand, preferably by spray, and the sand is continually stirred, usually by hand protected by rubber glove. A few trial runs will indicate how much sand to put in the sieves - normally about half the sieve floor should be covered - how much water to run through and how long to sieve a load. Each tray load may take ten minutes or so - it should be continued until the outflow water runs clear. Fines trapped in a tank may be saved for other experimental purposes and the organics are usually washed out as a scum.

4.2 Forming initial surfaces

Each experiment for demonstrating landform development must commence with some kind of initial surface. For some experiments this will be an isomorphous surface, and for others a scalar representation of a generalized real landform.

Isomorphous surfaces, suitable for demonstrating the commencement of rills, raindrop impact, the development of sinuous stream traces, desert ripples etc., may be produced by forcing an adjustable carriage over the sand along runners. In the case of flume work the carriage may be as in Figure 4.5. It consists of a flat board A spanning the trough. Along its leading edge is an upright blade B which may be adjusted to any height by means of bolts passing through slots C. The blade may be fitted with formers so that a variety of surface forms may be worked into the model surface. In Figure 4.5 a simple scribe is fitted to produce a V-groove ready for a stream experiment.

The carriage should be moved in one direction only, slowly scraping off excess sand and dumping it at the seaward end. The head pool is often retainable for replicate runs. While the carriage is being forced down the trough, the set-square member D should be kept firmly pressed against the flume wall. It is convenient, but not essential, to mount the carriage on rollers broad enough to ride along the flume edge without damaging it.

In order to produce horizontal beds, the flume walls themselves must be trued to horizontal to act as rails for the carriage. If inclined surfaces are needed there is a choice of tilting the flume, which is possibly the commonest technique, or of using wedge-shaped fitments (inclined runners) on top of the flume walls to give the required slope (Figure 4.6).

Figure 4.4. Apparatus for wet-sieving large quantities of sand.

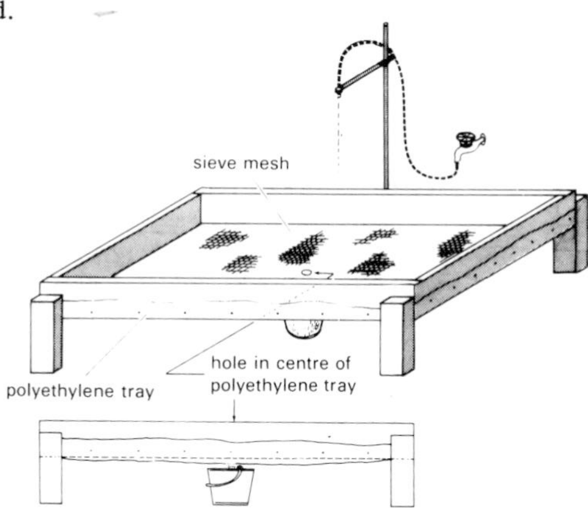

Figure 4.5. Adjustable carriage for producing isomorphous surfaces.

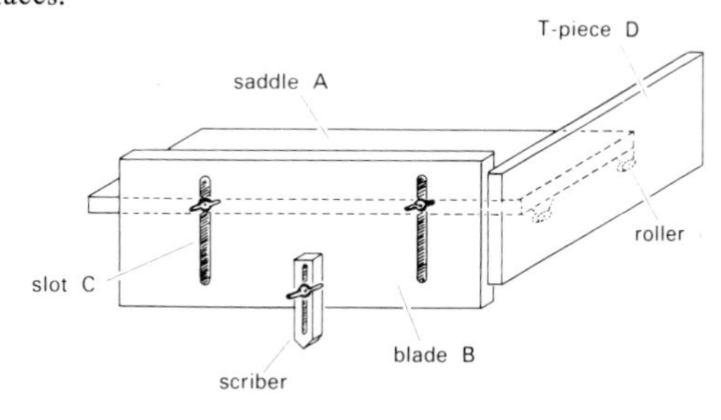

Figure 4.6. Wedges on flume wall for producing sloping test bed.

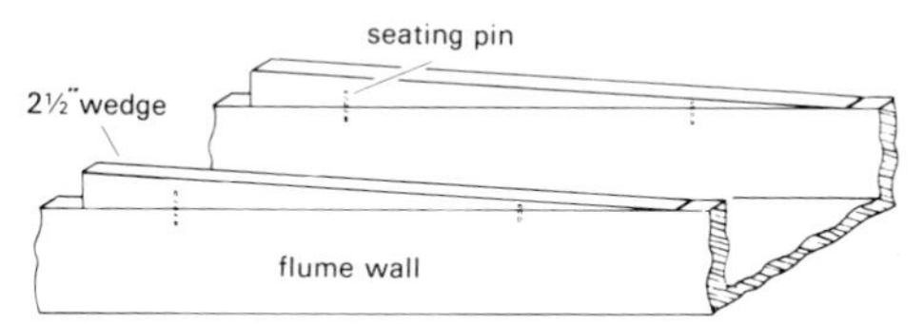

It is only the sand mass above the water surface (water table or base-level) that is of moment in this connection, and the water level will adjust to an equivalent position regardless of how the inclination is given to the sand surface. This is of some concern with regard to the use of small flumes or stream-tables where the ability to tilt often appears to be a prerequisite of construction, but is seen on reflection not always to be necessary.

If isomorphous surfaces of other types are required the construction of suitable formers is usually a matter of adaptation. A device found to be useful for producing smooth and accurate cones in the rainfall simulator is a pivoted sweep beam (Figure 4.7).

Beam B is pivoted on a centrally located stainless steel peg A in the trough. The beam is adjustable up or down a removable steel tube C located over peg A, so that the height of the cone and consequently its slope may be varied. The beam is hinged at the pivot end and fitted with a sleeve and tightening bolt. The outer end of the beam is fixed by a stay to the top of tube C, or for more rigidity to a ceiling shackle centred over the tube, the stay also being adjustable by tension screws.

When the model has a sea, coastal sand will settle at an angle of rest, and isolated pieces of the landmass will fall into the sea, establishing a static equilibrium condition. The presence of underwater coastal debris may be of no consequence, and indeed the sea itself might have a sand floor. Frequently, however, it occurs that erosional and depositional forms of micro-magnitude will require observation, and it is necessary that the sea bed should be cleared of pre-run sand. This may be done with the sea *in situ*, since the entry of the sea creates slump conditions. The clearing of the sea bed and trimming of coastal strips and shorelines is in this situation extremely difficult.

Figure 4.7. Pivoted sweep beam for producing smooth and accurate cones in rainfall simulator.

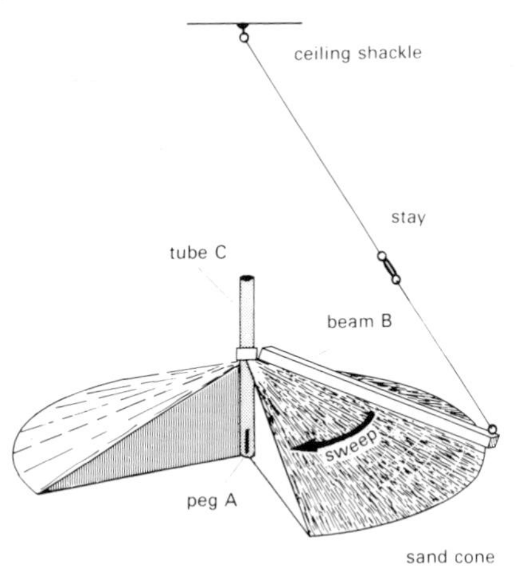

Brushes are not sufficiently controllable. A stiff rubber blade has been found to be effective and a square-edged sponge does a neat job, but whatever scraper is used it has to be moved through the water so slowly that turbulence does not dislodge new material.

An effective way of tidying up the coast is to run the sea in gently, to exactly the water level at which the model will run, and then to empty it out. The water/coast equilibrium form will have been achieved, and the removal of sea-water will not disturb the wet cohesive sand. A sponge may then be used to remove all unwanted debris from the coastal zone. Water soaks into the sponge and sand grains adhere to it at the same time. When the required finish is achieved, sea-water must be replaced with infinite care, so that no eddies or currents dislodge more material. Waste sand may be removed from the apparatus or it may conveniently be scraped out of the way to a side of the trough.

The working sand which is always lying about models during cleaning, formation and running must be prevented from entering the outlet pipe. It is often possible to use a lip or flange on the outlet, but this determines the last few millimetres of sea-depth. If a flush outlet is needed a piece of rubber tubing may be lain across the path to the outlet, and a few centimetres from it. In experiments where falling sea-level is used, the outlet pipe will usually be far too high above the trough floor to permit escape of particles. In no circumstances should the outlet pipe be protected with gauze since this will seriously affect freedom of flow, and errors of several millimetres water depth may be introduced.

The return of water into a mobile-bed model before the commencement of a run requires care. It is not usually sufficient simply to lie a hose in the trough and turn on the water, as this frequently produces currents which can wash out carefully finished sand edges. Some system of baffles is required to diffuse the flow, and one of the most effective is a plastic pot, perforated with sufficient 3.2 mm or 6.35 mm holes around its base, and filled with nylon pan-scrubbers. The inlet hose can be seated firmly in the centre and water leaves the pot with adequately dissipated force. In some cases it is necessary to syphon water into the model trough, since this offers an easily controlled, slow, eddy-free delivery.

4.3 Notes on general plumbing (assuming a re-circulating system)

One of the first requirements in any handling of water for model construction is an understanding of water flow in pipes and other equipment. The main disadvantage of small stream-tables used in simple hit-or-miss demonstration is their unsuitability from the plumbing aspect.

The friction of water in pipes is sufficient to slow down circulation, especially under gravity, to negligible rates, so piping should be of a wide diameter. This applies especially to the outlet pipe which should be at least 2.5 cm (1″) in diameter, and more if possible. 1.27 cm ($\frac{1}{2}$″) piping is almost always inadequate for proper control of sea-level. Water from the outlet pipe should also be free-falling – it should leave the apparatus and fall immediately into a tank (Figure 4.8).

If it is necessary to lead the water off from the apparatus to a distant point (a sink or tank) then attention should be given to the possibility of pumping the water out of the pipe, or at least of using a 3.8–5.1 cm internal diameter pipe, and bends should be avoided.

Generally, the apparatus can be designed to permit a free-fall outlet, and the vagaries of the re-circulation system are then related to water level in the catching tank, rather than sea-level in the apparatus. For this reason, the catching tank should be tall (Figure 4.9), but need not be of great volume. The height is needed to give hydrostatic pressure to force water to a main reservoir tank. Piping between the catching tank and the main reservoir tank must be of at least 3.8–5.1 cm internal diameter.

The reservoir tank (Figure 4.10) is an important item. Its volume must be calculated so that it will hold, if necessary, all the water in the complete system, including the sea from the model trough. It must be accessible for emptying and cleaning, which should be done every week or so with disinfectant. For this reason it is not convenient to use the reservoir tank for the receiving tank, since if used in that position it would be beneath the trough. If this seems to be for various reasons the most suitable position for the reservoir tank, the

Figure 4.8. Adjustable outlet pipe in stuffing box.

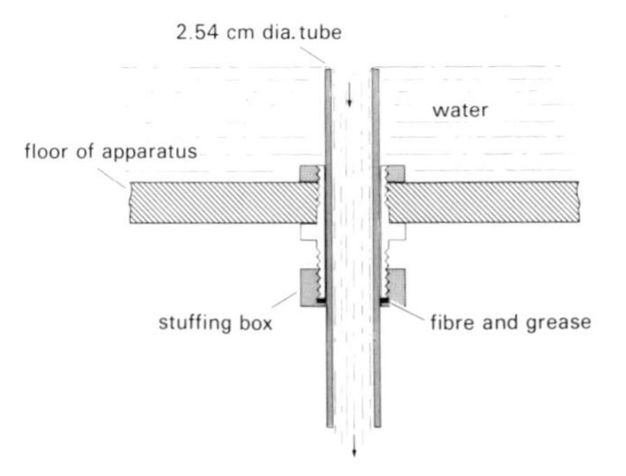

Figure 4.9. Requirements for typical catching tank.

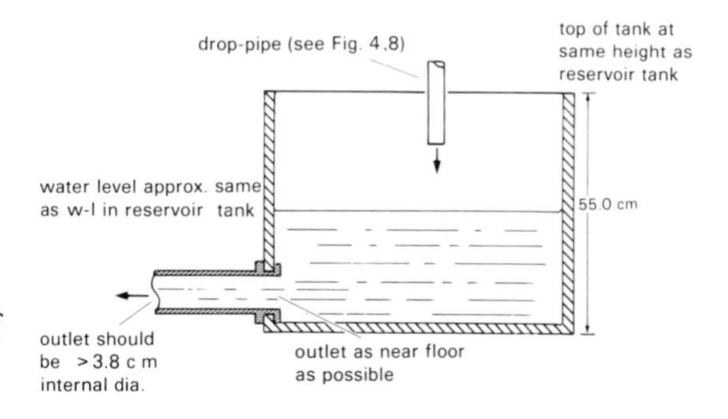

Figure 4.10. Requirements for typical reservoir tank.

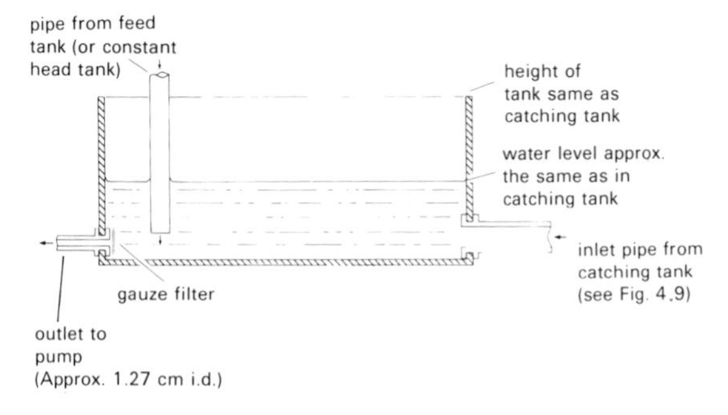

sea-outlet pipe will have to drop water straight into it, or be led into it by chute or flume after free-fall. Another problem with locating the large reservoir tank beneath the outlet pipe is the height restriction introduced by the apparatus floor.

If the reservoir tank is located towards the feed-system end of the apparatus, the large-diameter pipe connecting it to the catching tank must be free from upward bends, to prevent the formation of air-locks. This can happen while the system is being cleaned, or during a run if the reservoir tank is pumped too low. Ideally, the water should be able to move downhill from the catch tank to the reservoir tank through a pipe of constant inclination.

The reservoir tank will feed the pump from an outlet near its base. This will reduce the problem of the pump having to be primed after cleaning the system, or after inadvertent suction of air. There is no need for water under strong suction or pressure to be carried in large-diameter pipes, and 1.27 or 2.54 cm piping is quite adequate for most apparatus. It is necessary to have a filter mesh over this outlet from the reservoir tank to prevent foreign matter being sucked into the system.

There are several suitable types of pump for flumes and rainfall simulators. In general, the pump for a flume need not be as powerful as that for a rainfall simulator, since the pressures needed in flume delivery are less than those for rainfall simulation.

For either equipment a centrifugal pump has advantages. It is cheap, efficient, easy to control and does not readily develop faults. The impeller pump can be driven by a motor of $\frac{1}{4}$–1.0 h.p., depending upon whether the impeller requires very high rpm or not. For instance, in flume work a small pedestal pump run at 1410 rpm has been found to be adequate, while for rainfall simulation, a similar pump may need to be driven at 7000 rpm – or else a more powerful pump will be needed.

Usually the pump will have fitment sockets which give some idea of the size of piping to lead into and out of the unit. The pump will be gravity fed and will deliver water without difficulty to the height of the apparatus. Between the pump and the feed-system of the apparatus it is useful to have a cock and a pressure gauge. The flume gauge will be adequate at 9.8–14.4 g cm^{-2} (20–30 psi) and the simulator gauge should be capable of recording pressures up to 48 g cm^{-2} (100 psi). The cock should be the gate type and is fitted between pump and pressure gauge (Figure 4.11). This permits control of delivery, and pressure variation indicates certain types of fault that may develop during a run.

From the pump, water is supplied to the feed-system of the apparatus, and since the design of feed-systems is specific to flume or rainfall simulator, and is related to the type of phenomena to be studied in each apparatus, this is discussed in more detail in appropriate chapters.

The use of plastic piping is preferable to metal or rubber in general plumbing of apparatus. Joins should be properly cemented and plumbers' tape should be used with metal to plastic or metal to metal joints.

Ideally, apparatus should be waist high. This makes it easy to work with, permits suitable drain-off gradients to tanks located on the floor, and facilitates the use of benches and sinks at the same level.

Motors and other permanent apparatus should be firmly bolted to the floor or to the flume stand – but attention should be paid to avoiding vibration, which can impart a directional bias to vector components in the model.

4.4 Ancillary equipment

In addition to formers and tools for shaping and trimming models, a range of measuring devices is needed.

Rough vertical measures may be taken with rulers mounted on a lead-weighted base (Figure 4.12) which are placed on the trough floor near

Figure 4.11. Positioning of gate cock.

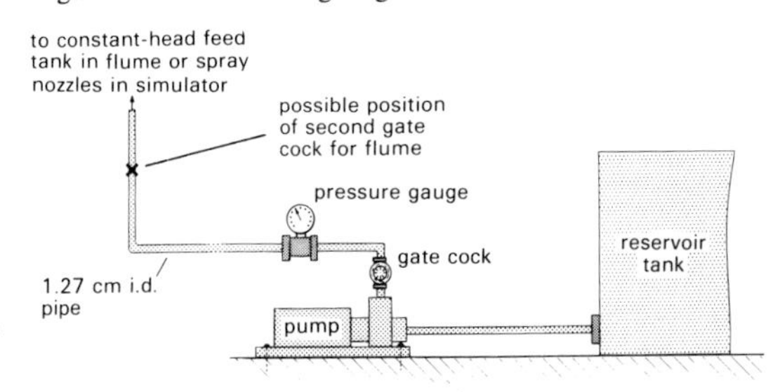

the model, and are useful for estimates of height during a run, or for photography. These also give a check on sea-level. Rough horizontal measurements are made with a metre-stick.

Accurate measurements are possible by means of sensing devices, and at some stage in model use a decision has to be made regarding accuracy and sophistication. This will concern the nature of the phenomena being studied, the effect upon understanding and conclusions associated with varying degrees of error, and the convenience and availability of measuring devices. When the same point was discussed with respect to major apparatus, a preference for simplicity was explained, and the same preference exists here. There is a natural tendency to develop more elaborate measuring devices and to use, for certain types of work, electronic or photo-electric instruments. However, consideration of the main model objectives shows that mechanical measuring is usually appropriate.

In dynamic geomorphic models the aim is to produce processes and forms the same as, or conceptually analogous to, larger-scale natural phenomena. Some dynamic studies, and perhaps most, will be satisfied by lower measurements in the centimetre and millimetre range. Occasionally this may be inadequate, as in accurately measuring sheetflow, but while the demonstration of the possible thinness of films of water is within the framework of the present discussion, it is perhaps an uncommon exercise.

Measurement of relief in centimetres and millimetres can be effected by use of the azimuth bar, which is of particular value in the rainfall simulator. It is a ⊥ shaped unit (Figure 4.13) suspended from the centre of the apparatus and perfectly balanced by hair-weights at either end, so that the cross-beam is horizontal.

The horizontal member has a datum line between its two parallel bars, spring-loaded to a high tension. These bars are graduated from the centre in metres and centimetres, and the whole can be rotated through 360°.

The bar is used in association with a vertical probe graduated in centimetres and millimetres, which is lowered gently through the horizontal member of the azimuth bar, just in contact with the datum line. If the probe is first used to get a reading on sea-level, all other landform heights in the experiment can be adjusted to this.

Also, as the bar is rotated and the probe is slid along from one position to another, horizontal distances from the centre of the trough to the nearest 0.5 cm are obtainable.

Readings to the nearest millimetre, vertically, are quite easily obtained in this way, and, with

Figure 4.12. Simple depth- or height-gauge.

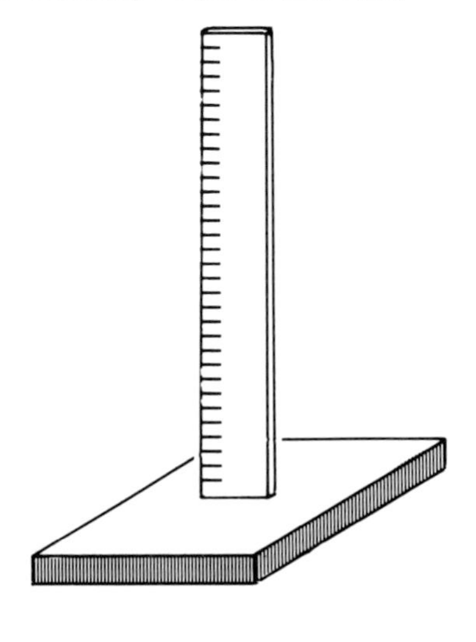

Figure 4.13. Azimuth bar.

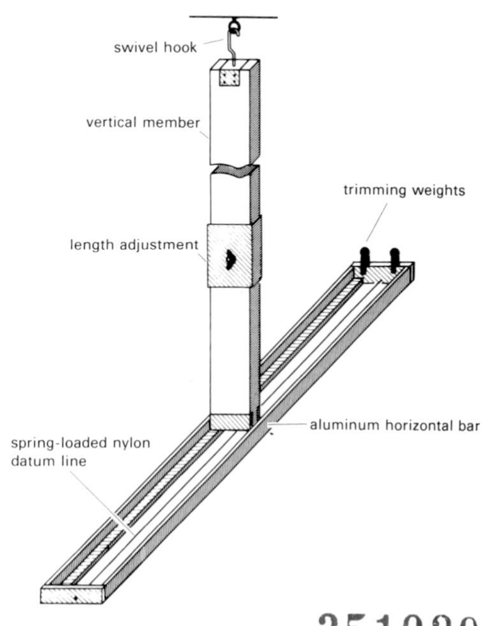

practice, 0.5 mm may be assessed. This technique builds up a series of points in a three-dimensional graticule, using a central point and an arbitrary line as initial references, but this translates into sea-level and some fixed point on the model. The data may readily be converted to a relief map or block diagram (see Figure 6.1).

Horizontal measurements to finer units than those offered by the azimuth bar may be achieved by using small metric rules, preferably transparent and of thin section, tapering to a calibrated knife-edge.

A simple electrical device for measuring water surfaces such as sea-level, stream surface gradients or groundwater level, is a fine metal probe and an ohm-meter (Figure 4.14). The meter should be set at $\times$ 100 Ω and the probe should be insulated along its shaft, excepting only the lower tip. This is very sensitive, but should be read only during moment of contact, since at withdrawal the probe draws up a surface-tension thread which gives considerable error.

If it is used (possibly in conjunction with the azimuth bar) for determining depth of the water table, perforated tubes should be pre-sunk into the model mass. These may be of brass or plastic with large-diameter holes drilled as shown (Figure 4.15). To prevent the entry of sand they should be wrapped in fine gauze, and their bases must be plugged. They must be inserted in the model so that their lowest holes are below the expected water table.

When the probe is lowered into the tube, the contact tip must be kept away from the side of the tube, and several readings should be taken in each location to guard against inadvertent false contacts.

In flumes, the most convenient way of obtaining linear measurements in three dimensions is to fit a rule to the lower edge of the sand former - or to a separate carriage (Figure 4.16). One edge of the flume should be calibrated, and with a probe held against the horizontal carriage edge all dimensions are measurable. This is essentially the system described by Allen.[6]

In some models it is possible to place transparent overlays in position on the model surface, so that features may be traced off. Measurements may then be taken from the overlay.

Photography is useful for recording form and process sequences, and may be used later for measurement, but in general photography will not play a role in routine demonstration.

A small plumb-line has frequent uses in conjunction with the measuring devices described.

Slopes are often best determined by taking horizontal and vertical measures and plotting them out on graph paper. If standard slopes are to be used for certain models, a device such as that shown in Figure 4.17 may be used. A spirit level gives the horizontal, and the slope arm and upright are capable of being locked by a screw to any angle from 5° to 20°, and from 20° to 40° in larger units.

Velocity of sand movement has to be determined by time and linear measure. Velocity of

Figure 4.14. Simple electrical device for measuring water levels within sand model.

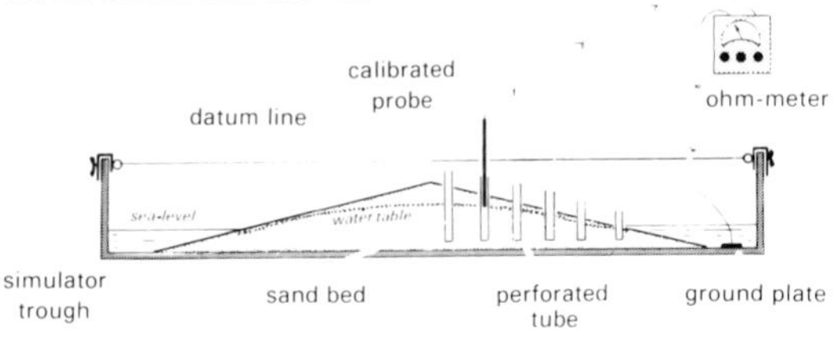

Figure 4.15. Perforated tube for subsurface water-level probing.

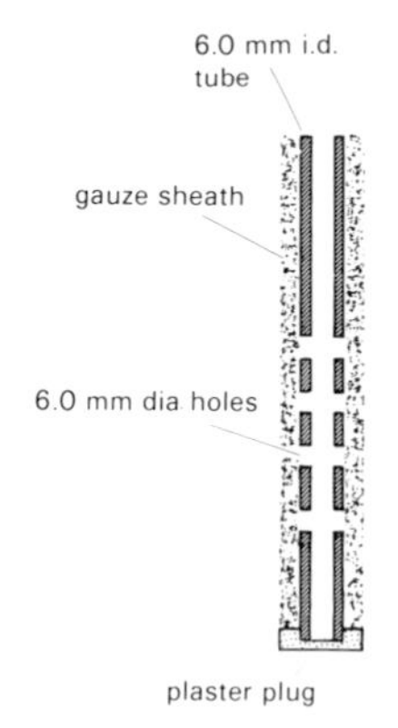

Figure 4.16. Rule fitted to edge of former blade for lateral measurements.

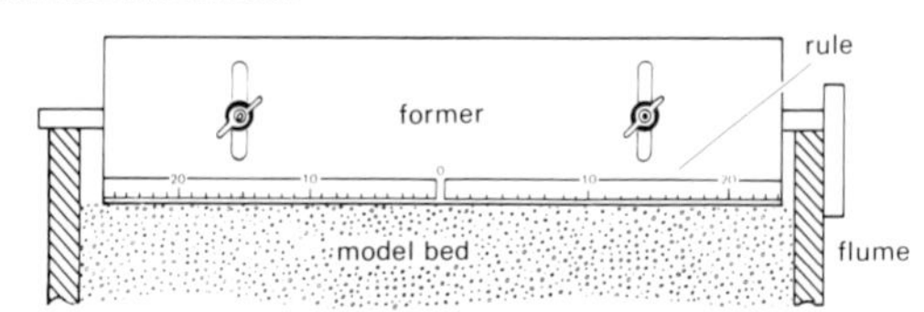

water flow in rills, streams and overland is most conveniently measured by introduction of a dye such as red ink. If flow is laminar, stop-watch timing is taken from the moment the dye hits the water until the first tendril of coloured streamline passes a datum downslope.

Dyes have the disadvantage of colouring water in a re-circulating system, and usually require flushing of the total system at the termination of an experiment.

Micro-floats, such as granulated sawdust or particles of polystyrene, are affected by surface tension and tend to be attracted to the sides of streams or coastlines. Also, they are difficult to remove from sand at the end of an experiment.

Pitot tubes are of value in larger streams,[7] but the readiest method for most purposes is probably small flecks of potassium permanganate, which give a clear reading but do not seriously discolour water.

Generally, channels in models will be too small in section to permit insertion of a pitometer, but mean velocities may be accurately calculated by catching the free-fall output from the trough - that is the outlet from the sea - and transferring it to a measuring cylinder. Velocity in the channel will then be

$$\bar{V} = \frac{Q}{\bar{d}wt},$$

where Q = total volume of catch, $\bar{d}$ = mean depth of channel, w = width, t = time in seconds for catch Q.

Since $\bar{d}$ and w may be taken at any section in the channel or equivalent marine section, this is a versatile technique.

The production of raindrops of a size that will not create splash erosion may be achieved by trial and error. With spray systems, as described in Chapter 5, this should present no particular problem. There are, however, good instructional reasons for knowing the size of raindrops, and an easy technique is the flour-pellet method of Laws and Parsons.[8] Providing the droplets are sufficiently small, a layer of plain flour can be spread on a microscope slide, and this can be passed through the rainfall. Any flour wetted by a drop will form a sphere of the same diameter as the drop, which can be measured under the low-power microscope against a millimetre scale.

Rainfall intensity should also be known since it is a critical factor controlling to some extent the rate of erosion. A suitable technique is to lie petri dishes in a pattern over the trough and to catch a 5-10 minute sample. The catches are carefully transferred to suitable measuring cylinders. If only a coarse value is required, a standard raingauge may be set up in the trough, but this may not take a representative sample.

Figure 4.17. Device for measuring slopes.

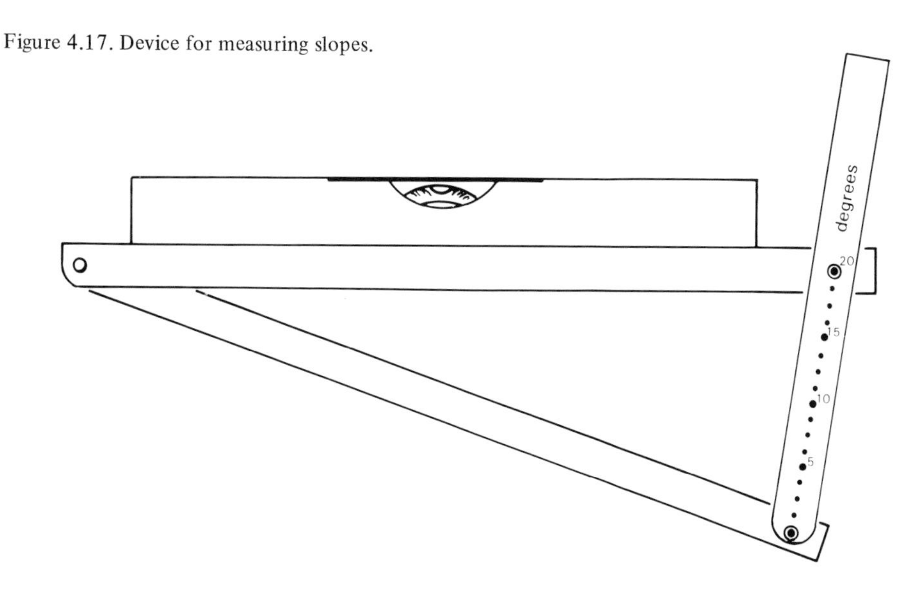

Sediment is usually measured from points of deposition around the edge of a landmass. This is the deltaic or slump debris. If the floor of the flume or simulator is smooth, as it should be, the debris can be removed by rubber scraper. Alternatively, a small sheet of polythene can be laid carefully in the sea at the river mouth or cliff foot before the run is commenced. It may have to be weighted down at the corners.

The sheet and sediment should be removed while the sea is still in position. Sediment may be dried before weighing by leaving it spread thinly on newspaper to dry in air, or, preferably, in a drying oven. The most convenient balance is a triple-beam type with a scoop pan. As sediment samples tend to be of >100 g, finer balances are not normally needed.

Other measurements, such as feed into the flume and sea output, may be affected with beakers and measuring cylinders, and thought should be given to accessibility during equipment design and layout.

5 The rainfall simulator in demonstration

The commonest dynamic models in the laboratory fall broadly into categories of rivers fed by point source, coastal features using a wave generator as in Chapter 9, and models of a general erosional type depending on simulated rainfall.

Occasionally, reference is made in model literature to the use of a water sprinkler over a stream-table. This is an ineffective procedure that should be avoided as self-defeating. The need is for water of suitable drop size to fall with uniform, natural velocity and distribution over the model surface, and this requires a certain amount of careful equipment design.

5.1 The principles of rainfall simulators

The critical aspect of simulators is the manner in which drops are produced and delivered to the model surface, and simulators are classified on this basis. Some methods of producing raindrops are novel, and a good summary is given by Pearson and Martin,[1] but the three commonest methods are: dripping water from threads projecting through the base of a reservoir supported above the test surface; dripping water from spaced capillary tubes; and spraying water over the test surface by nozzles.

Most of the work in the use of rainfall simulators has been carried out for agricultural purposes, and Hayward[2] has reviewed 102 studies involving this type of equipment, while Mutchler and Herschmeier[3] have compared the three main delivery systems. Of the three, the hanging yarn method is least adaptable for drop size, although it can give a good, if coarse textured, distribution. The capillary tube method is much more adaptable and gives a non-random, uniform coverage, but is also coarse, unless the sprinkler system and model surface are moved in relation to each other. Capillary sprinkling has been described by many workers[4,5] and is used in the versatile Illinois simulator which may be programmed to give a wide range of storm patterns.[6] These cases are not approached from the demonstration viewpoint, but illustrate design techniques and use potential.

Spraying usually requires a fairly high nozzle pressure. The technique takes several forms and is versatile as regards drop size and area coverage, but is not capable of highly uniform precipitation-distribution. This method of rainfall production has been discussed by a number of workers,[7,8,9] and a spray device was used by Würm in 1936 for some classic work in demonstration.[10]

Rainfall simulators are becoming common in various university and government laboratories for erosional and hydrological research. They are often quite small, especially if indoors, when dimensions of 100 cm × 50 cm in plan would cover some instruments, but outdoors they tend to be larger and are often modular or readily transportable, so that large test plots can be covered. If the drop system is yarn or capillary tube for laboratory use, a modular system is often preferred so that the apparatus may be extended.

The yarn and capillary systems are relatively inflexible, and it is not easy to produce an overhead reservoir tank into which tubes or yarn can be properly sealed. The yarn technique is cheap, but drop size is practically uncontrollable. Capillary tubes can be designed for different drop sizes. The drops in all these cases are large and their terminal velocity is too low unless the delivery system is suspended high above the model.

Aspects of reality that have to be borne in mind in rainfall simulation are intensity, drop size and distribution.

5.1.1 Rainfall intensity. If the model is to be used to demonstrate scaled reality intensity will have to be known, along with other precipitation parameters. In rainfall simulation input to the model system may need to be greater than the prototype input. Thus, although normal intensities can be used, better effects may be obtained with higher intensities. Records of natural rain from many countries indicate that while in temperate climates precipitation rarely exceeds 75 mm/hr, in tropical climates it often exceeds 150 mm/hr.[11] Intensity values in Britain are commonly in the 2.0–10.0 mm/hr range, while 60 mm/hr is uncommon.[12] In tropical conditions analysis of 113 storm intensities showed a range of 0.56–127.0 mm/hr, with a majority of less than 10.0 mm/hr. In simulator work, higher intensities are usual (e.g. Anderson[13] 203 mm/hr, Yen and Chow[14] 89 and 170 mm/hr, Bisal[15] 102 mm/hr). The reason is that, as in using unconsolidated sediment for bed material in river or coastal models to give a foreshortened time scale, so by increasing precipitation intensity in surface-erosion studies, greater activity in shorter time can be achieved. This does not appear to produce serious process problems of the kind that would result if very much more powerful waves or river discharges were used in models.

In a simulator designed for dynamic model demonstration an intensity of over 12 mm/hr but not in excess of 40 mm/hr is normally advisable, since at less than 12 mm/hr the effects of rainfall will probably be too slow, and at over 40 mm/hr they may be too rapid.* Also, at low intensities it is much more difficult to control drop size and distribution.

5.1.2 Drop size. Simulators for dynamic model demonstration will need to be capable of producing small-diameter drops. In this they will differ from simulators to be used only for process-exploring experiments. It is customary in research to use drops of a size that occur most commonly in local rainstorms. The upper limit of natural drops is usually taken as 5.0 mm, but the velocity and diameter of drops up to 6.1 mm diameter have been measured.[16] In agricultural and soil studies generally, drops used have varied from 0.3 to 6.0 mm.[17,18] Such large drops are conveniently produced by the yarn or capillary tube technique, and the result of large drops falling on carefully formed micro-surfaces is cataclysmic. Although the terminal velocity of these large drops may be less than in nature, their energy, which is a function of mass and velocity,[19] will not allow fine erosional features to develop. When rainfall is required to fall upon a surface without splash, to infiltrate, develop a water table and produce run-off with resulting microforms, a small drop size must be used. Drop sizes of mean diameter of about 0.2 mm are recommended, with a maximum of no more than 0.5 mm in a given storm. Even this will produce some sand splash, but not enough to destroy micro-landform features. If these drops are allowed to fall through about 1.5 m, the terminal velocity of the larger drops will be between 1.0 and 2.0 m s^{-1}, and the kinetic energy of most drops will be too small to dislodge a sand particle. A fall of 1.5 m is much less than that commonly used in soil erosion research in which terminal velocities of 7 to 17 m s^{-1} are required.[20]

Drop size may be checked by a modification of the flour-pellet test,[21] and other ideas may be gained from Pearson and Martin's summary.[22]

5.1.3 Distribution of rainfall. The use of capillary tubes and yarn permits uniform distribution of rainfall, and if the construction of the system is modular, considerable extension of distribution is possible. For dynamic models in demonstration it is not normally necessary to bring a storm in and then phase it out, and with fine sprays this is not practicable. The main problem in the sprays is to maintain a uniform distribution over the test surface. A single spray nozzle will produce different intensities within its own wetting area, and if more than one nozzle is used, it is difficult to balance areas of overlap although, with care, such a balance can be achieved.

5.2 Construction of a simulator capable of producing required rainfall characteristics

The size will be determined in part by available laboratory space and cost, but consideration should also be given to observation potential. The larger the simulator, the greater the number of students who may closely observe processes, and an effort should be made to have this piece of

* Certain experiments require much higher intensities, see Experiment 1, this chapter.

apparatus, together with a flume, as standard equipment in earth-science and physical geography laboratories.

The details of construction given will concentrate upon essentials so that they can be developed for simulators of various sizes. The optimum considered here is based upon one constructed at Trinity College, Dublin and later duplicated, with minor modifications, at Memorial University, Newfoundland. Such a model may as easily be used for research as for demonstration, and this interchangeability of goal is characteristic of well-designed apparatus.

The key part of the system is the spray, and by far the most suitable has been found to be a brass nozzle marketed as a tree-spraying head, normally fitted to a long lance (Figure 5.1).

Other types of spray devices have been tested, and it may be taken that sprinklers using a simple gravity feed through a perforated plate or nozzle head (i.e. the watering-can principle) will be of no use in this model context. Drops will be too large, and very likely poorly distributed and unsteadily delivered. Smaller drops require a pressure feed of some kind, and tap-supplies may be adequate. The various delivery systems marketed, such as spinners, vibrators, broken-jet devices, fish-tail distributors etc., are designed to work off tap pressure, but usually have large drops and poor distribution patterns - and intensity is almost impossible to control accurately. It is only with the pressure nozzle with a single small orifice that suitable characteristics of precipitation are obtainable.

The majority of spraying devices, including the nozzles, produce a coarse spray containing large drops of water that fall in a heterogeneous pattern. This pattern is usually steady for any given pressure or nozzle setting, and areas of heavy and light density only shift their locality if these controls are changed. A further problem with many delivery systems, including nozzles otherwise suitable for this work, is leakiness around the head, or condensation beads that drop upon the model surface, pitting it and spoiling micro-effects.

Figure 5.1. Tudor nozzle: a readily obtained adjustable spray-nozzle.

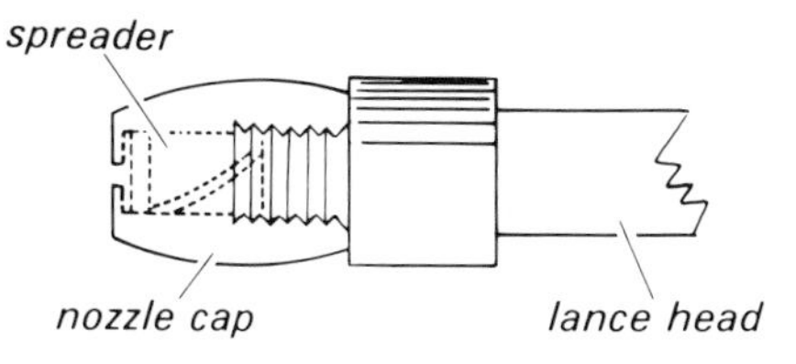

The Tudor-type nozzle adjusts by screwing the cap onto the lance head, and it is sealed against leakage under high pressure by a rubber collar seated in a machined groove. Within the nozzle is a nylon spreader which fits snugly into the end of the lance. As water passes through the lance and around the spreader it is given a rotational movement by two spiral grooves in the spreader sides. This rotation is not detectable in the spray. Spray characteristics are changed by variations in water pressure between the cap and the spreader, manipulated by adjustment of the cap.

The ex-factory nozzle will require careful finishing, possibly under a low-power microscope or lens. The orifice will have a burr on the inside edges and on the spreader - which in the preferred nozzle is nylon, but in most nozzles is brass - and the orifice is not usually of exact diameter. The burr may be removed with very fine emery or alumina paste, and the orifice must be trimmed to a uniform size by means of a tapering needle which is forced into the orifice up to a mark on the needle at a micrometer-determined diameter. The needle may be used to standardize a set of nozzles for combined systems.

The nozzle is then tested by rigging it to a falling-head permeameter or funnel and rubber tube system, that will deliver an exact quantity of water with a uniform initial head of pressure over a measured interval (Figure 5.2).

Figure 5.2. Falling-head 'permeameter'.

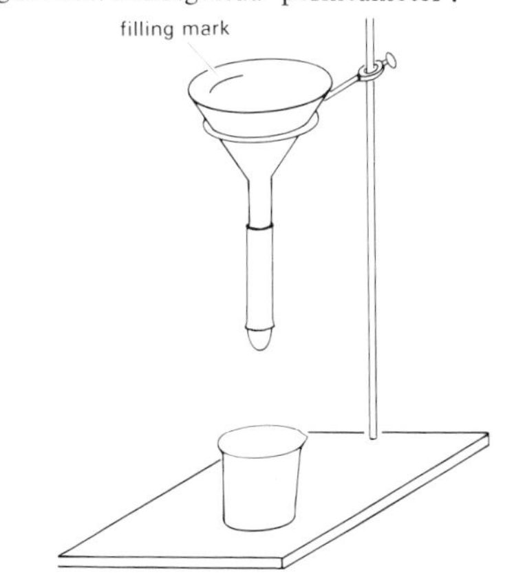

The most effective orifice diameter at pressures just capable of forming a fine spray is about 1.59 mm ($\frac{1}{16}$″). At 3.17 mm ($\frac{1}{8}$″) diameter, drop size is too large and control over distribution is lost.

The spray cone is parabolic (Figure 5.3) with a spread of about 1.1 m in a 1.1 m fall, and beyond that distance spread barely increases. The nozzle settings tend to produce spray phases. If the nozzle is opened very slightly from zero, an extremely fine mist is produced. Upon further opening the mist form is maintained, with increasing intensity, but the parabolic spray cone becomes ill-defined. At a certain critical point of opening the delivery changes to a much more sharply defined cone of small diameter, but of small drop size, and this gives way on further opening to a spray of large, high-velocity, pressure-forced drops, covering a small surface and capable of dislodging sand particles. The most satisfactory delivery settings are therefore closer to zero than full-opening.

This type of nozzle has on occasion been fitted to a laboratory tap under ordinary water-mains pressure, to be used over small stream-tables (Figure 5.4).

Figure 5.3. Parabolic spray cones with different nozzle settings.

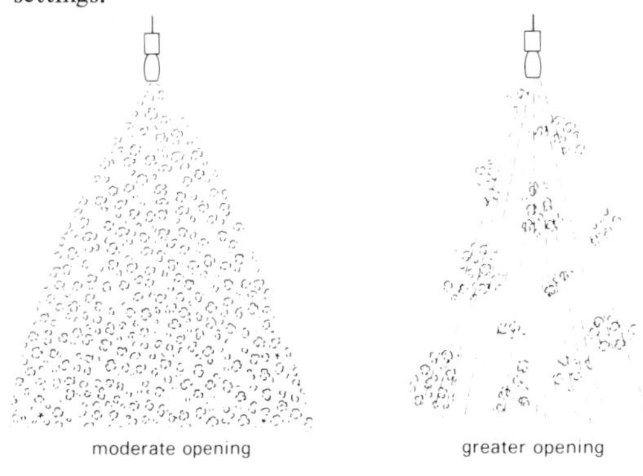

Figure 5.4. A simple spray system using tap-pressure.

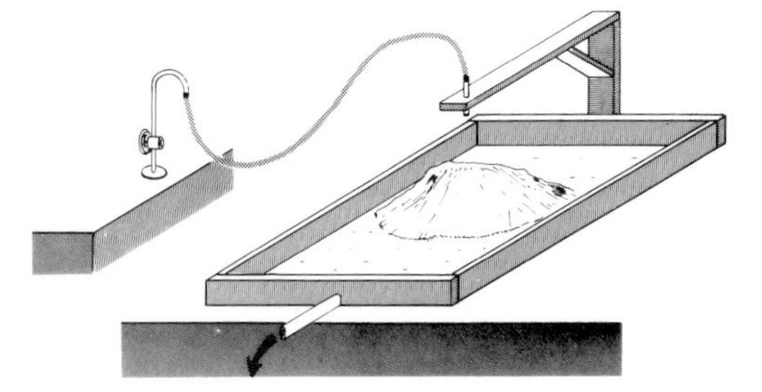

Table 5.1. *Precipitation distribution with five nozzles*

central pan	30.0 ml
20 cm radius	20.0 ml
40 cm radius	1.5 ml
60 cm radius	pans collect a few microdrops
80 cm radius	pans clouded or clear

In large simulators, more than one nozzle is usually needed and they will have to be balanced. After calibrating with the falling-head feed-system (Figure 5.2), which can give orifice accuracies to within $\pm<1.0\%$ of the mean, the nozzles are put into the simulator system one at a time and their delivery patterns are determined.

Rainfall intensity and distribution are measured by a pattern of raingauges on the simulator floor, or, if the work is to be very accurate, upon a mock-up of the model to be used. A single nozzle, with an orifice of about 1.59 mm ($\frac{1}{16}$″), and with a feed pressure of 4.06 kg cm^{-2} (58 psi), delivering 7.6 ml s^{-1} over a six minute period with a fall of 1.1 m, will show a distribution in concentric circles as shown in Table 5.1. There will be a degree of asymmetry, the characteristic property of a given nozzle, which cannot be rectified without sophisticated machining techniques.

If more than one nozzle is used, it will be necessary to adjust them all to the correct delivery balance, and then to arrange them in the delivery system in such a way that they blend their wetting areas. Theoretically, with a distribution of 30/20/15 ml trace in concentric circles, a position exists in which two nozzles can be placed to give a line of values 30/>20/>20/30 ml, and if the model is placed in the *c.* 20 ml zone it will receive a quite uniform distribution. As an aid to producing a good distribution, nozzle patterns can be plotted in histogram form. Nozzles may be shifted and individual intensities slightly varied by cap adjustment until histograms are acceptably uniform (Figure 5.5). Since this is one of the most troublesome aspects of setting up an apparatus, once a good distribution is attained it should be left alone, only periodically being re-tested by petri-dish analysis. The adjustment of overall intensity by pump-pressure control should not seriously affect the distribution pattern. The nozzles are best fed by a re-circulating system driven by a centrifugal pump as in Figure 4.11. For economy, a small pedestal pump can be driven at high rpm by a step-up pulley train (shielded) to give maximum pressure of about 4.2 kg cm^2 (60 psi). At any setting pressure should be capable of being held steady by means of a gate cock in the delivery piping (Figure 5.6). It will be found that with the nozzle type described, satisfactory spray conditions commence at about 1.4 kg cm^2 (20 psi), and that below this pressure distribution is difficult to control. With increase in pressure above 1.4 kg cm^2 (20 psi) distribution characteristics of precipitation on the test surface improve up to about 2.8 kg cm^2 (40 psi) when they once more diminish. However, if the nozzles have been carefully prepared and the pump runs steadily, none of the distributions should be too poor for model running, and when necessary considerably higher pressures can be used.

The feed-system described will produce raindrops of approximately the correct size with an average of about 0.2 mm diameter. This will vary slightly with rainfall intensity in a manner indicated in Table 5.2.

Note that in using the flour-pellet technique for determining raindrop size the slide should be passed rapidly through the rain close to the model surface.

This spray system may be adapted to a small simulator trough of 1 m square, or to a larger unit suitable for a wide variety of models. A description will be given for a larger, more flexible unit, but the essential principles will be applicable to most other simulators used in this way, and scaling down is possible.

Figure 5.5. Sketch from laboratory notes of histogram method used for building up preliminary spray pattern from beam-spaced nozzles.

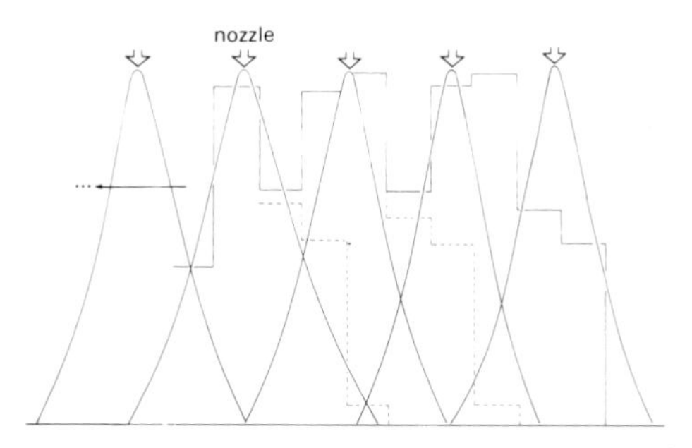

Table 5.2. *Variation of drop size with intensity*

I (mm/hr)	Maximum diameter (mm)	Mean diameter (mm)	Minimum diameter (mm)
14.4	0.5	0.21	0.05
17.6	0.4	0.20	0.1
20.8	0.3	0.21	0.1
24.0	0.3	0.18	0.1

The simulator consists of a trough for the model, a super-structure for the rainfall apparatus, and a pumping system. If the trough is large enough, it can double for a wave tank and for a flume for certain types of experiment. An ideal trough has been found to measure 2.44 m (8.0 ft) along each side by 0.3 m (1.0 ft). It can be made of 0.95 cm ($\frac{3}{8}''$) pre-waterproofed plywood screwed into a floor frame of two-by-fours (5 cm x 10 cm). A lighter frame may be used, but attention has to be paid to the base-support. This may be of concrete blocks or bricks, and will be designed to give solid support to the trough.

When the floor has been constructed, sides are screwed into it and strengthened with corner uprights (Figure 5.7). The trough's interior should be sanded, caulked and painted white. It may be lined with 1000 gauge polythene sheet which is available in suitable widths, care being taken not to fold the plastic into a sharp corner since at this thickness it tends to produce a hair crack. An alternative waterproofing is fibreglass, which, although requiring more effort, is the ideal finish, being entirely waterproof, tough and durable. However, polythene linings have been widely used in a variety of situations and have survived years of experimentation with a minimum of attention. If the sheeting is handled carefully during initial construction, common-sense usage, particularly in avoiding contact with sharp instruments, will ensure lack of problems.

The trough is raised far enough above ground level for easy working over the upper edge, which should be about two feet high.

Figure 5.6. Reservoir tank and pump assembly for rainfall simulator.

Figure 5.7. General construction of rainfall simulator.

Above the trough, a framework of wood or angle-iron is constructed to carry the spray system and side curtains.

For a 2.44 m square trough, five or more nozzles will be required, arranged according to the test area, intensities and model possibilities. The nozzles should be about 1.2 m above the possible model surfaces, taking into account lack of parabolic spread at greater distances, and terminal velocity of small water droplets. The frame should be strong enough to carry a system of water-filled piping, nozzles and brackets, curtains and such ancillary apparatus as lighting and measuring instruments. A metal framework is therefore recommended.

The nozzles are fed by a tube system which keeps the individual distances from main feed pipe to each nozzle as nearly equal as possible. If one nozzle is remote it will tend to receive water under lower pressure than the others and give tuning problems.

If a very good distribution of rainfall is necessary, some type of moving nozzle system will have to be employed. This may take the form of a travelling beam which moves back and forth on rails, driven by an endless chain. An example of this system is shown in Figure 5.8, in which the beam is 7.5 cm x 10.0 cm (3″ x 4″) hanging from a carriage at each end. The carriages are fitted with ball-bearing rollers which travel upon polished and greased 2.5 cm (1″) angle-iron rails. The chain should be a roller chain of at least 0.94 cm, with a key link carrying a projecting pin which engages a greased slot in the beam carriages (Figure 5.9). As the chain travels it draws the beam with it to the end of the trough, at which point the chain loops around a sprocket and the pin reverses direction, drawing the beam back again. A sufficient drive for this system is a $\frac{1}{4}$ h.p. motor reduced from 1425 rpm to a beam speed of 52 cm s^{-1} (1.7 ft s^{-1}) which is fast enough for the beam to return before raindrops have settled from the previous sweep (Figure 5.10).

Figure 5.8. Sweep-beam with five nozzles and feed-pipe.

Hose in the pressure system should be heavy-duty, braided, 1.25 cm inside diameter, and should be flexible enough to move with the beam.

Water in the trough should lie at a uniform depth, which requires careful floor levelling. The exit is through a free-fall pipe at least 2.5 cm (1″) inside diameter. The pipe should be as flush with the trough floor as possible, but has to be attached by a flanged fitment so that the polythene lining of the floor can be clamped down around the hole (Figure 5.11). Water moves into a catching reservoir via a 5 cm (2″) inside diameter pipe. This tank may be small, but its rim should be above the height of any sea-level likely to be used in the main trough (Figure 5.6). If the outlet pipe is not wide enough or has too many bends or obstructions, control over sea-level in the trough is difficult.

When water has fallen into the catching reservoir it moves under hydrostatic pressure into a main storage tank, which is large enough to contain all the sea-water from the experimental trough. With a re-circulating system and a situation in which considerable volumes of water will be constantly fed into and emptied from the experimental trough, a rapid containing device is needed, unless a floor drain is conveniently to hand. Pumping water into sinks between experiments is slow and inconvenient.

From the reservoir tank water passes by gravity through a 3.75-5.0 cm ($1\frac{1}{2}$-2″) internal diameter pipe to the pump at floor level. The inlet to this pipe should be covered with fine gauze to

Figure 5.9. Detail of beam carriage with drive-chain.

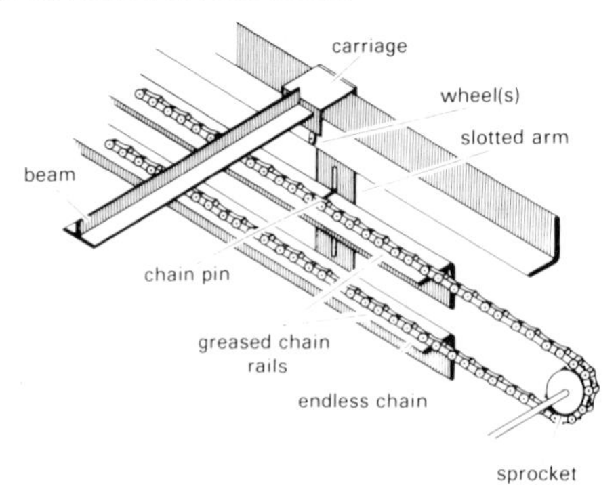

Figure 5.10. Drive-shafts, sprockets and gutters at ends of beam traverse.

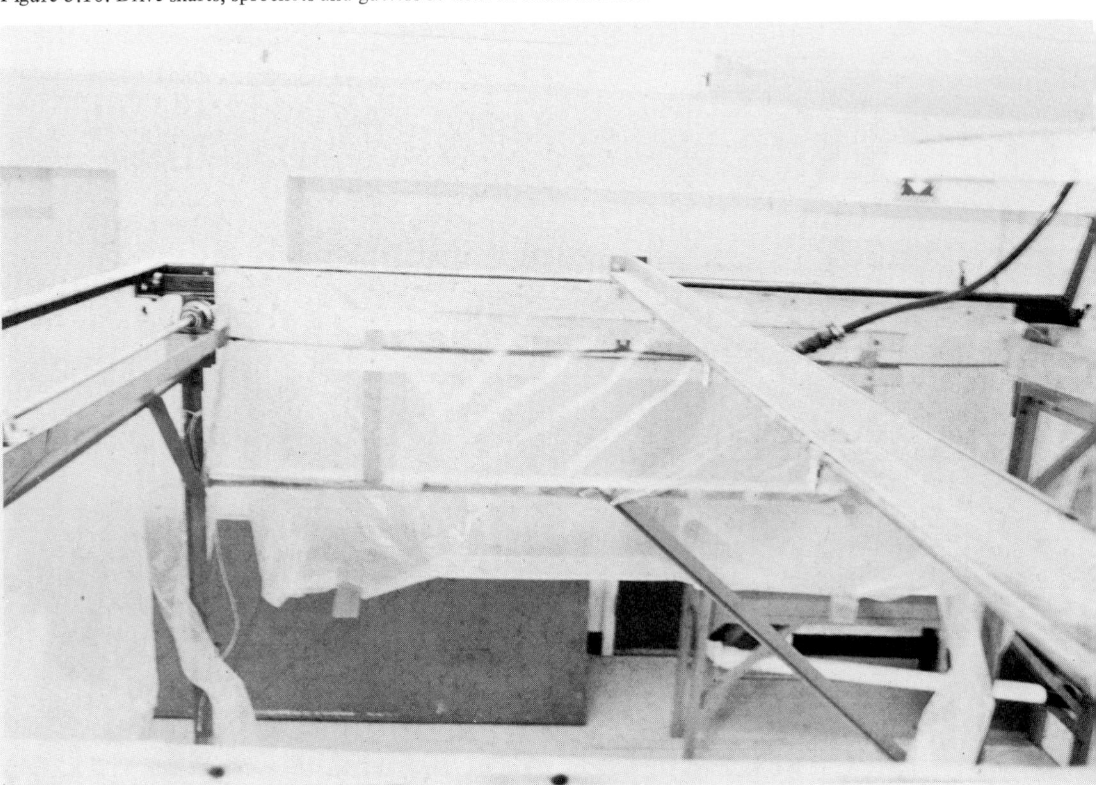

prevent passage of foreign bodies, which will eventually lodge in spray nozzle orifices, requiring them to be stripped down, cleaned and re-tuned.

Switches and electrical controls must be mounted away from the simulator trough or other places where water is likely to exist.

Curtains are of 500 gauge polythene and may be tacked onto the superstructure cross-bars. When the apparatus is running, curtains mist over and it is necessary to cut out viewing flaps of about 5 cm (2″) square, leaving a hinge at the top. Any number of these may be cut and they can be prevented from opening outwards by paper clips attached as in Figure 5.12.

5.3 Problems of operation

Calibration and tuning of the simulator involve the feed-system and the experimental trough. The nozzles are the most difficult part of the apparatus, and even after careful cleaning and adjustment they require regular checks, since it is usually impossible to assess by eye whether a spray is delivering below design quantities. After initial construction, or after repairs or adjustments to the feed-system, the piping to the nozzles should be well flushed out by running without nozzle caps to rid the pipes of bits of rubber, metal or grit. The commonest problem in the feed-system is grit working along piping from the reservoir tank, sucked in by the pump which, being centrifugal, can pass it on to the nozzles.

Figure 5.11. Flanged and bevelled outlet with large diameter return pipe.

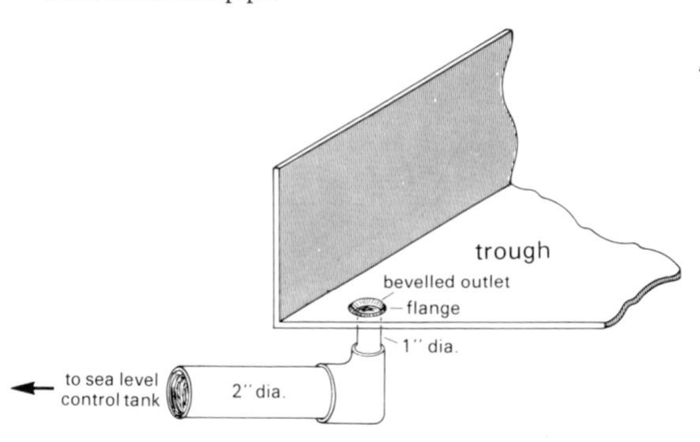

Figure 5.12. Viewing flap in simulator curtain, showing attachment of paper clips.

Delivery can also be affected by variations in electric current – this was found to be particularly problematical in the tropics. Constant checks on the pressure gauge (Figure 5.6) give a good monitoring method, and for convenience the gauge can be re-marked in intensity values, with emphasized markings at any commonly used pressures.

Water in the reservoir tank should be maintained so that when the experimental trough is carrying sea-water (which has probably been pumped from the reservoir tank to begin with), enough water is left behind to prevent a suction vortex developing over the pump intake. It is not uncommon for air to be sucked into the pump even though water level in the reservoir is 5.0 cm (2″) above the outlet pipe. This spasmodically reduces pressure and causes spitting from the nozzles.

The outlet pipe from the experimental trough to the collecting tank should be seated as flush with the floor of the tank as possible, to facilitate drawing off remnants of sea-water without recourse to sponging or paddling water into the outlet. This pipe can cause problems of sea-level control since water tends to build up over a too-narrow opening. Care should be taken to prevent sand from falling into the outlet pipe, as it will increase pipe friction and inhibit flow of water.

Temperature changes should be noted. If the apparatus is in a building that is allowed to cool down at night, to say 11 °C, then it will be necessary to warm the water in the reservoir tank before an experiment can be run. The water could be maintained at a temperature of about 18 °C by a simple aquarium immersion heater. Ordinary tap water is also usually colder than ambient laboratory temperatures and it may be necessary to let it warm up before running an experiment.

In re-circulating systems, there is a tendency to leave the water for several weeks if it is not discoloured by dyes or dirty sand. This permits the development of bacteria and slime on the walls of the apparatus, and in hot climates other organisms colonize the system. It is advisable from both health and technical points of view to empty out the whole system after a day or two of use, and to disinfect occasionally with a non-film producing, non-dispersant chemical.

5.4 Experiments in the simulator

The dynamic models capable of being run in a rainfall simulator fall into four general categories:

(1) rainfall-dependent true-surface processes and microforms,
(2) rainfall-dependent analogue forms (and possibly processes),
(3) water-table dynamics,
(4) composite rainfall and point discharge experiments.

In order that the student shall appreciate the difference between true and analogue situations, attention has to be paid to the observation of microprocesses for their own sake, and alternatively to microprocesses as analogues of a larger-scale reality.

The simulator is commonly used in such a way that natural conditions obtain, excepting rainfall which is controlled. When used in soil studies for agricultural purposes, for example, this is its experimental role. When, alternatively, the simulator is used to model a larger catchment in runoff studies, as, for instance, with such an instrument as the Illinois simulator, then its role is to investigate reality by study of an analogue, in which dimensions and materials differ from those of the prototype.

Simulator use in demonstration experiments may take either form. If the simulator is set up to demonstrate the formation of micro-rills, then it is being used to sample reality, and demonstration can include data collection which relates to extra-laboratory rills of like dimension. If the simulator is being used to produce microforms that in some way resemble prototypes, but in magnitude and material differ from them, the simulator is being used to demonstrate reality through analogue conditions. In this case, analogues produced for demonstration are not usually constructed in such a way that definitive conclusions relating to reality may be drawn – the objective is to shift attention from force-time-quantity parameters to form-analogue or process-analogue. It is important that this duality of purpose in demonstration models should be appreciated and made part of any exercise, in order that the student should not feel intellectually disoriented, even though a given analogue or set of controls can be criticized in themselves – that is to say, criticism should be primarily concerned with detail rather than concept.

Water-table experiments are a tangential development of simulator use, and the fourth category, composite experiments, depending upon an energy source other than precipitation, introduces complexity.

(1) Rainfall-dependent true-surface processes and forms, in which geometric, kinematic and dynamic similitude are 1:1.

These include models which may be samples taken from the field to demonstrate (*a*) erosion, (*b*) deposition, (*c*) rainsplash, (*d*) sheetwash and effect of vegetation on runoff, (*e*) infiltration, eluviation and illuviation. Many of these can be based on soil research papers such as those of Bryan.[23]

5.4.1 Erosion. A tray or pan containing a sample of soil is set in the simulator, subjected to precipitation, and the material washed out is gathered in a trap and weighed. The experiment should be closely observed and data should be collected. As with any dynamic model or experiment, the controlling parameter(s) should be known, induced or deduced, and changes in the system should be recorded. Any data or information it is necessary for students to have should be tabulated and a formal laboratory-sheet should be prepared in some format such as the following:

Experiment: 'X' Date: 16 February 1979

Purpose: To demonstrate the quantity and nature of erosion on a (local) soil under high intensity rainfall.
Apparatus: Rainfall simulator, container a cm x b cm x c cm, raingauge, timer, etc.
Procedure: Undisturbed soil sample taken from Y (method of sampling can itself be recorded) and placed in container in simulator. After subjection to rainfall, the sediment eroded was removed, dried, weighed, sieved, (etc.).
Controls: Soil pre-wetted to field capacity
precipitation intensity________________
drop size________________(If required, drops of >1.0 mm can be produced by opening nozzles up to a firm spray.)
surface slope________________
duration________________
soil characteristics________________
Observations and measurements: Usually mass, energy, time and linear parameters together with cause-and-effect relationships: include time taken for initiation of surface storage; for initiation of erosion; appearance of runoff water (turbid, muddy ...); weight of dry filtrate; weight of wet residue; weight of dry residue; possible surface erosion forms.
Conclusion: This may be a systematic statement of process and results or it may be transformed to a systems framework. The essential factors will be that the soil, under the control conditions specified, is eroded at the rate of____________ g ml^{-1} of rain, or energy equivalent. Due to the fact that air resistance on fine drops is likely to produce a terminal velocity of less than theoretical v, in the equation kinetic energy $(E_k) = \frac{1}{2} mv^2$, the alternative equation potential energy $(E_p) = mgh$ may be used to satisfy the energy input factor, where m = mass in g of rain that has fallen on the *model* surface, g = acceleration due to gravity in cm s^{-2}, h = height of nozzle above model in cm and $mgh \times 10^{-7}$ = energy in J. Also, a conclusion may be reached concerning general erosion process and the composition of grain sizes of eroded material.

Experiment 1

Purpose: To demonstrate the quantity and nature of erosion of a soil under high-intensity rainfall.
Apparatus: A rainfall simulator with travelling beam, spray-nozzle delivery; soil tray 20 cm x 20 cm x 10 cm surrounded by 10 cm wide splash-in/splash-out zone and runoff slot (Figure 5.13); raingauge; timer; various glassware and auxillary equipment such as drying oven and balances.
Procedure: The soil was carefully removed from a field location, placed in the soil tray and moderately compacted. It was wetted to field capacity and the tray was placed in the simulator trough. Precipitation intensity was pre-set at a higher than natural value and was monitored during the run. Runoff from the soil tray was collected in a bottle and later measured for liquid and solid content.
Controls: Soil–sandy loam; angle of slope of tray surface 10°; rainfall intensity 315 mm/hr; drop size modal at 0.8 mm diameter; drop-fall distance 1.66 m; duration of run 2.98 minutes. Limit to downward erosion set by lip of soil tray.
Observation and measurements: Runoff catch 133 ml; sediment in 133 ml water 0.398 g; surface storage and runoff occurred within 30 s of commencement of run, murky water falling from initiation of runoff into catch-bottle below tray. The washed-off material was of fine sand and lesser grades.
Conclusion: When high-intensity precipitation falls upon a bare sandy loam soil at field capacity, saturation is rapid and surface storage occurs within a few seconds of commencement of rain. Within thirty seconds surface runoff occurs and fines are carried off immediately. During a 2.98 minute storm the weight of soil removed from a 400 cm^2 test surface was 0.398 g, which represents about 8 g/hr, or 200 g/hr over a 1 m square surface. As fines are the primary grain-size involved in this process it would be expected that the quantity of eroded sediment would diminish with continued rainfall, since the lip of the soil tray prevents removal of a thick soil layer and larger particles.

Note that the experiment has obvious limitations and further experiments immediately suggest themselves. The decision as to whether new experiments should be conducted on a given theme depends upon the initial core concept identified, and the level of development of the course.

The above information describes the essential aspects of a soil-erosion situation and gives quantities so that orders of magnitude can be perceived by the student. A more comprehensive, but tighter analysis, with a better presentation of cause-and-effect relationships is obtained by using the systems approach.

Conclusion (Systems approach): Since the container is a box from which the only exit is the open top, soil contained in it cannot escape from the soil mass except over the surface. This is more significant if the soil is at field capacity before commencement of rainfall, so that initial eluviation is reduced.

Given these conditions, the morphological subsystem is strictly speaking the air–soil interface, which is more or less irregular, with soil aggregate micro-relief and an overall slope of 10°. The particle size of this surface may be important and a sample could be sieved. In the sample used here, the soil is sandy with a small percentage of fines, but particle size has not been given status in the morphological subsystem.

The cascading subsystem is represented by a mass input of 626 g raindrop water, which has an impact energy of 10.19 J (kinetic energy) together with a runoff potential energy of 0.02 J, and a dislodged soil particle mass of at least 0.398 g with at least 1.4×10^{-5} J potential energy. Output from the system is divided into runoff and infiltration. Runoff output includes water mass (133 g), eroded soil particles (0.398 g) and a total kinetic energy of less than 10.21 J together with heat energy. The infiltration system output into the soil body is unknown other than the water mass, which is 493 g (Figure 5.14). Note (i) no account is taken of small amount of evaporation from soil surface during 2.98 minutes of run. Note (ii) input energy (J) = mass (g) of precipitation × distance (cm) of fall/10 197.16 = (626 × 166)/10 197.16 = 10.19 J + mass of water in runoff × distance of drop in elevation = (626 × 0.3472)/10 197.16 = 0.02 J (where $g = 980$ cm s^{-2} and is taken into account as a constant in 10 197.16).

The canonical diagram shows that the soil surface system has an input of water mass and

Figure 5.13. Soil tray, as used in Experiment 1.

energy, and an output of water mass, soil mass and energy. In considering such a system it is necessary to indicate that kinetic energy will be converted to heat, but that measurement of heat energy would require more complex instrumentation. Soil erosion, as it is generally conceived, accounts for the measured output of water and soil. The output to the subsurface system is not directly monitored but 493 g of water is derived by deduction.

In the soil surface system, a potential positive-feedback loop exists in which decreasing infiltration capacity increases surface storage, which increases surface runoff and increases soil dislodgement and erosion. In the field, this could be expected to decrease infiltration capacity further by infilling of pore-spaces at the surface, thus increasing runoff and erosion. In the laboratory model a limit to erosion is set by the tray walls, so that surface storage will increase as the soil surface is lowered, until ponding effectively inhibits further soil loss. This condition was not reached in the 2.98 minute running time.

If the course is at an introductory level then student involvement may be simply that of visual assessment. There is evidently a range of possible treatments, and the treatment could be more intensive than that outlined above. It could include collection of data on porosity of soil, moisture content before and after experiment, formal grain-size analysis of field sample etc. For further sophistication the experiment could be repeated using different rainfall intensities, slopes, durations and so on, and results could be subjected to statistical analysis.

Figure 5.14. Cascading subsystem for Experiment 1.

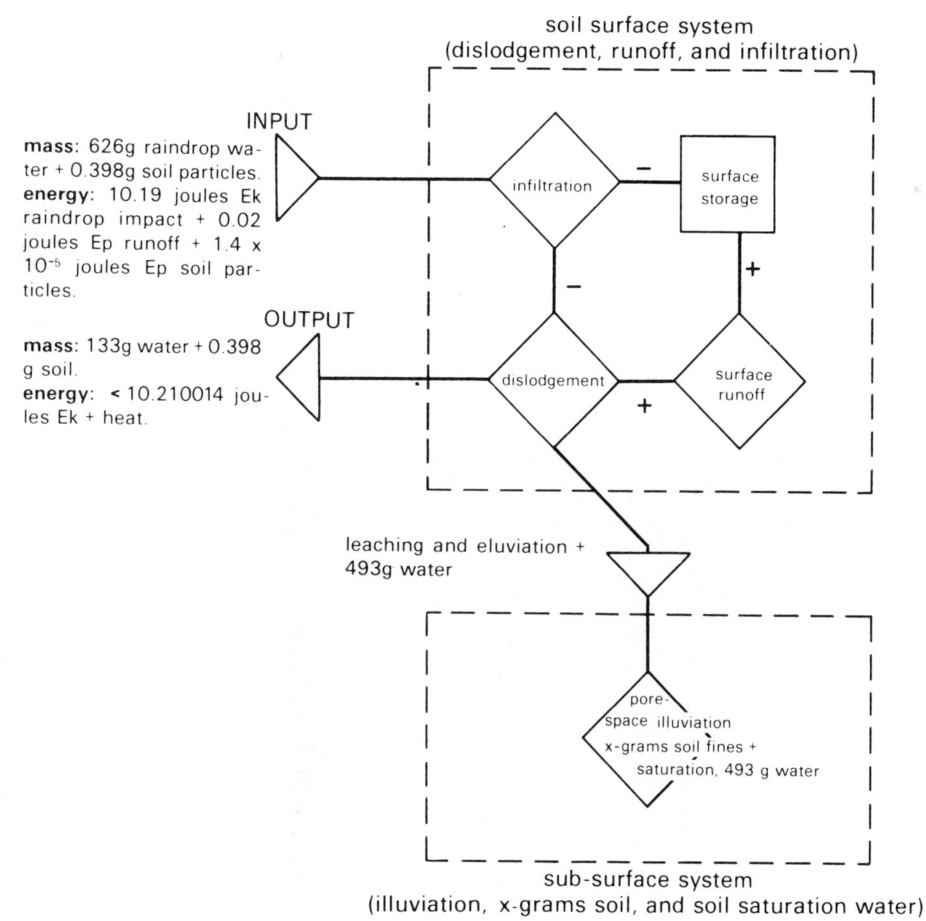

If it is felt that more comprehensive treatment is required, it may be that rather than demonstration, which is instructor-guided, a minor research project could be given, since with this type of model there is no sharp line between demonstration and research.

For other experiments concerned with demonstration of real processes, a similar format can be adopted. Each experiment should be regarded as a natural system in which certain factors are controlled in the laboratory, attention should be focused on the essential phenomenon, and data collected should apply to very specific earth-science principles.

5.4.2 Deposition. In so far as this refers to the re-employment of an eroded and transported sediment, it could involve micro-conditions such as occur in soil surface depressions. A depression can be produced in the laboratory on a natural soil surface of known parameters and can be covered with a layer of fine sand so that material deposited may be identified by contrast or section at the end of the run. Providing the sand does not retard infiltration in the depression, sediment accumulated should be by natural deposition.

Experiment 2

Purpose: To investigate the nature of deposition of materials resulting from surface wash in a declivity in soil.

Apparatus: Prepared soil; rainfall simulator to produce moderate-intensity precipitation, soil tray (Figure 5.15), timer, raingauge and various glassware and auxiliary apparatus.

Procedure: The rainfall intensity was pre-set but monitored during the run with a raingauge. A plastic tray of 40 cm × 25 cm was filled to a depth of 4.5 cm with a natural soil passed through a 16 mm sieve to remove larger stones. With the base of a porcelain evaporating dish a circular depression was formed 12.5 cm diameter × 1.0 cm deep, care

Figure 5.15. Soil tray, as used in Experiment 2, showing declivity in soil.

being taken to avoid destroying absorptive capacity of surface. The declivity was then covered with a thin layer of fine sand, by direct sieving over a template. The tray was then placed in the simulator and subjected to rainfall. After the run, the model was removed and an assessment of deposition was made by inspection of surface and section through the declivity.

Controls: Soil <16 mm diameter; precipitation intensity 18 mm/hr; drop size modal at 0.8 mm diameter; angle of tilt of tray 2°; duration of run 5.17 minutes.

Observations and measurements: The sequence of events was as follows. At MINUTE 1.83, surface storage commenced; at MINUTE 2.5, storage in main depression started; at MINUTE 4.5, sediment began to wash into the hollow; at MINUTE 5.0, sedimentation was accompanied by micro-rilling in the sides of the declivity; at MINUTE 5.17, the run ended.

From post-run model surface it was seen that the material washed into the depression was of fine grade and derived from the sides of the depression and the general flat surface above. Sedimentation followed erosion on the downslope side of the depression, where the soil was saturated first (and in fact puddling occurred on the soil surface at the downslope end of the trough). Inwashed fines covered about 11% of the 123 cm^2 hollow, giving an area of deposition of some 34 cm^2. At about 1 mm thick this gives *c.* 9.0 g of fines deposited in 5.17 minutes.

Conclusion: Under the control conditions used, depressions in soil will act as collecting centres for surface storage and eroded fines from adjacent areas. In the experiment carried out an area of less than 500 cm^2 produced deposition over 34 cm^2 of the depression surface in a total storm period of 5.17 minutes.

The systems interpretation for this example of depression deposition is essentially the same as that for erosion, in which output from the erosion surface will equal input to the depression system. This will be a negative-feedback (self-destroying) system, since deposition will eventually modify the morphological subsystem (the depression) to the extent that it will cease to exist as an effective *draw* for surface flow.

5.4.2 Rainsplash. This phenomenon offers a number of experimental possibilities. Splash-out from a tray of soil can be monitored to give a general assessment, and, if required, intensity, drop size and distance of fall can be varied. Another possibility is to study the effect of one large drop falling upon a soil surface, and yet another is to study the effect of a sequence of large drops falling upon the same spot. The use of a rainfall simulator is not required in every case.

Experiment 3

Purpose: To study the effect of a single raindrop falling upon a soil surface.

Apparatus: A firm vertical rod of retort-stand diameter and about eight feet in length; clamps and retort rings (Figure 5.16); optional use of a cantilevered sample-carrier (Figure 5.17); a drop-producing device consisting of a gravity-feed from a suspended funnel releasing water through a piece

Figure 5.16. Apparatus for studying effect of a single raindrop falling on soil surface, as used in Experiment 3.

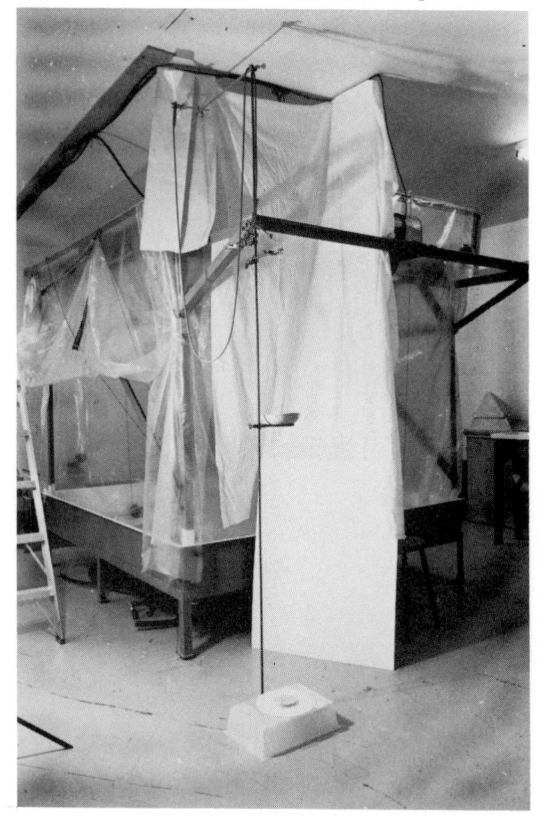

Table 5.3. *Rainsplash data*

	Fall distance (cm)	Inner diameter of crater (mm)	Outer diameter of crater (mm)	Splash-out globules		
				No. on sand	No. on filter paper	Total
A	243.5	17.0	20.1	4	11	15
B	179.5	16.5	18.5	10	7	17
C	128.5	15.0	17.0	8	9	17
D	13.5	13.0	Not discernible	0	0	0

of glass-tubing drawn to a narrow neck, joined to the funnel with rubber tubing and controlled by a pinch cock (Figure 5.16); a petri dish; filter-paper; glassware and microscope.

Procedure: Silica sand was sieved into a petri dish and the surface was smoothed by drawing a straight-edge across it. The dish was then placed upon a large filter-paper on an inverted tray beneath the drop-producing device (Figure 5.16). The drops were pre-set and were falling at a rate of one drop per two or three seconds, and were prevented from hitting the model-surface by an interposed, filter-paper lined evaporating dish. The height of the drop-tube was pre-set and measured before each experiment. At a chosen moment, the shielding evaporation dish was removed, one drop was allowed to fall on the soil surface, and the shielding dish was re-interposed.

Figure 5.17. Cantilevered sample-carrier.

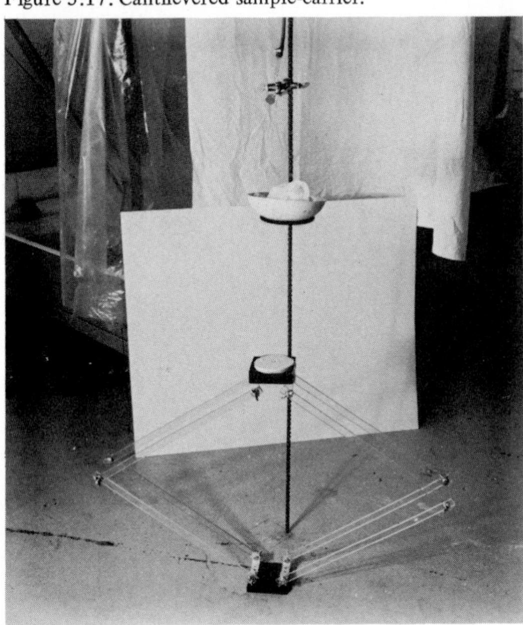

The resulting impact phenomena were then studied beneath a low-power microscope, and measurements were taken. This was repeated for four different heights.

Controls: Sand <0.35 mm diameter; drop size constant at 4.5 mm diameter; distance of fall 243.5, 179.5, 128.5, 13.5 cm.

Observations and measurements: Upon impact the drop produced a sharply-defined crater, ejecting spheres of a sand–water mix, the sand being derived from the crater. Within the crater a residual accretion mass of sand lay near the centre, which beneath the microscope showed as water–sand tendrils reaching out across the base of the crater. There was a small raised lip of sand around the edge of the crater (Figure 5.18).

Measurements taken were as shown in Table 5.3.

Conclusion: The measurements indicate that crater diameter is closely related to drop-fall distance (Table 5.3) and, although no measurement of crater depth was taken, the craters were of spherical form and it may be deduced that volume of soil dislodged is also directly proportional to drop-fall distance (Figure 5.19). The dislodged material is partly ejected from the crater, frequently over six centimetres distance, and partly falls back into the centre of the crater as an amoeba-shaped mass. The volume of this crater debris is more or less proportional to crater size. Globules from greater drop-fall-distance impact tend to travel further, but the number of globules produced does not seem to be directly related to drop-fall distance. There is a complex relationship between energy of drop at impact, particle size and condition of soil (sand), and number and size of globules produced.

The erosive power of the raindrop, under the control conditions used, is clearly demonstrated. *Systems interpretation;* The morphological system is the crater developed and ejecta produced in a very brief time interval upon a fine-textured smooth sand surface, which is not necessarily an analogue form or material.

The cascading subsystem is the mass of the single raindrop at and immediately following impact, with its kinetic energy, together with the mass of dislodged sand and its energy.

In so far as a single event changes the sand surface from plane to pitted, a micro-scale catastrophe has occurred. The forms studied after the event are merely a residual morphology which, during examination, are affected by a subsequent cascading subsystem of dessication and debris collapse.

The mass input of water is 0.048 g, with kinetic energy at impact of 1.15×10^{-3} J, together with dislodged sand of x g mass and y J kinetic

Figure 5.18. Four single-drop craters from varying drop-fall distances from Experiment 3.

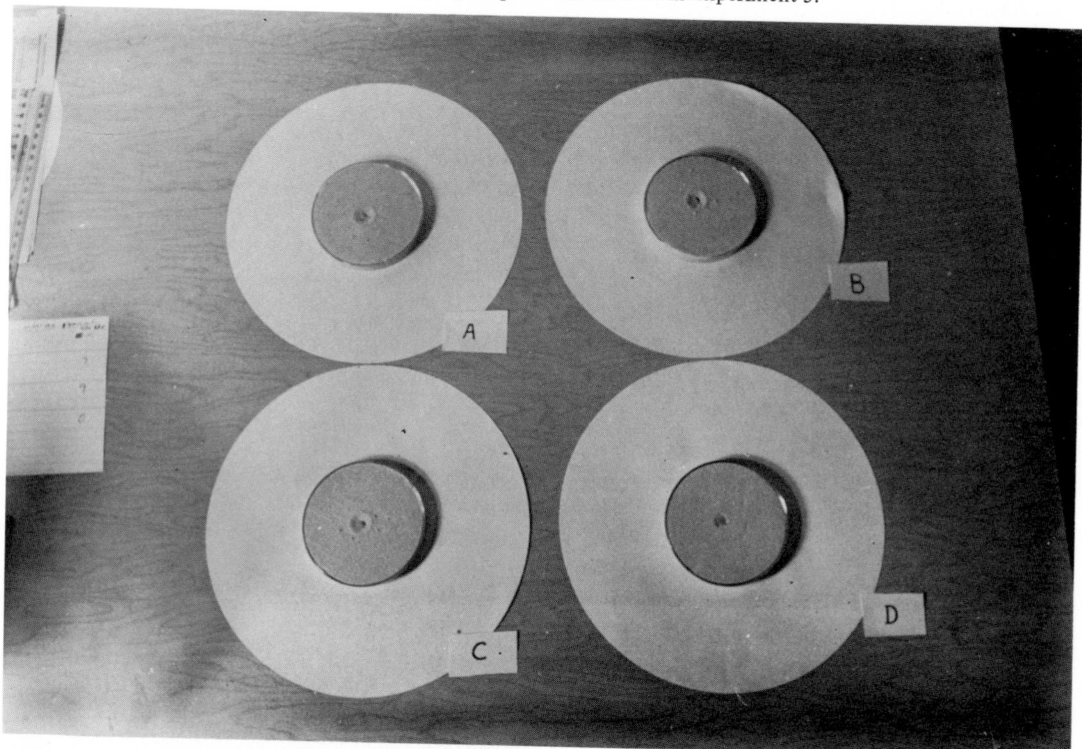

Figure 5.19. Data from Experiment 3, showing relation of crater diameter to drop-fall distance.

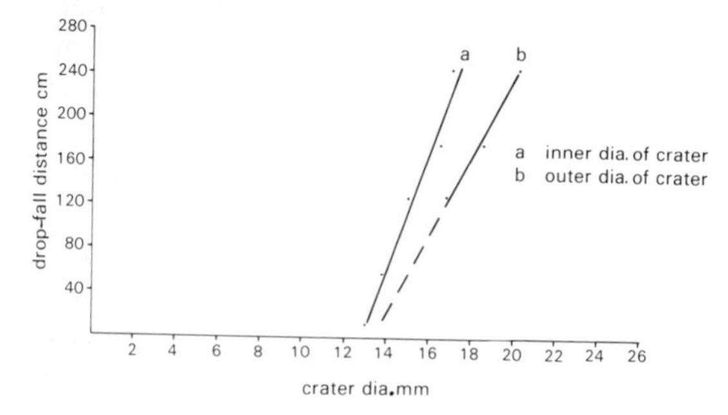

energy. Output from the crater is 0.048 g water, distributed as ejected water or crater-bed retained water, with less than 1.15×10^{-3} J kinetic energy, together with less than x g sand (since some returns to the crater) with less than y J kinetic energy, plus heat energy (Figure 5.20).

The energy of moving sand (y J) is derived from raindrop energy. Potential energy of the system is that of the position of particles in flight, with an optimum in each case when the ejected particle is at the apex of its trajectory.

Experiment 4

A modification of this experiment is to permit a sequence of drops to fall upon the same spot to increase erosive effect and, if the soil surface is sloping, a demonstration can be given of the commonly described phenomenon of raindrop-impact on a slope.

Purpose: To determine the effect of a sequence of raindrops falling upon a given spot on a soil slope.

Apparatus: A drop-producing device, as in experiment 3, together with ancillary apparatus including a timer; a block of medium sand dampened into cohesiveness (21 cm x 17 cm x 4 cm thick) which rests upon a stand to permit tilting (Figure 5.21).

Procedure: The block of thoroughly wetted and compacted sand was formed on a supporting base, smoothed at the top and the edges trimmed. This was placed on a stand beneath the drop mechanism. Drops were pre-set for size and frequency but were prevented from falling upon the test surface until time for commencement by a removable intercepting dish. The precipitation was commenced, timed and stopped, and the result was then assessed by inspection of the new surface and dissection through the crater (Figure 5.22).

Controls: Sand modal at *c.* 0.25 mm (see appendix: *standard laboratory sand*); drop size 4.5 mm; drop-fall distance 40 cm; drop frequency one per 4.5 s; sand surface slope 14°; duration of run 25 minutes; number of drops *c.* 330.

Figure 5.20. Canonical diagram for crater system in a planar surface (Experiment 3).

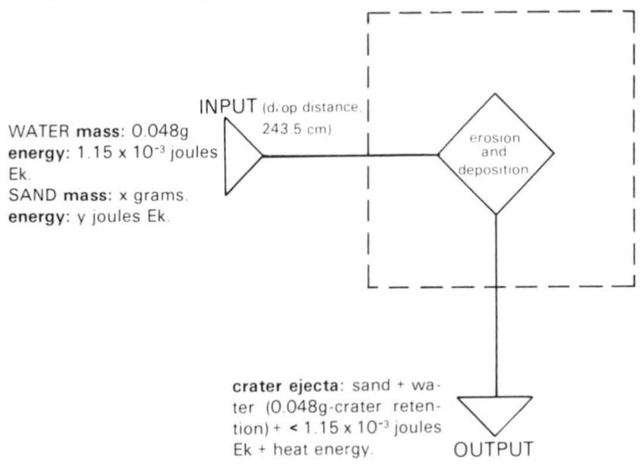

Figure 5.21. Tilt-board used for Experiment 4.

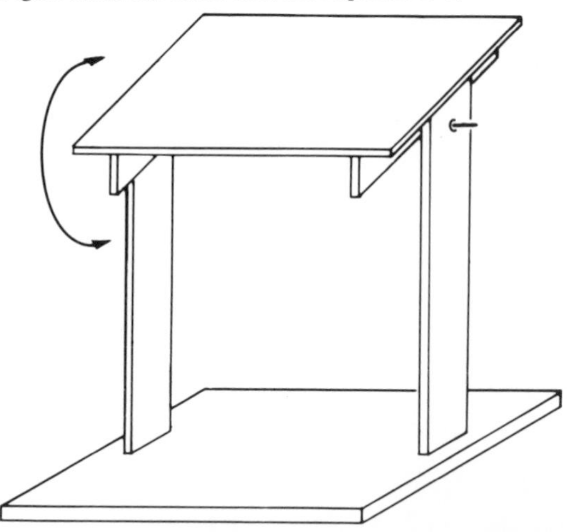

Figure 5.22. Sand block from Experiment 4 resting on tilt-board.

Observations and measurements: As each drop struck the sand surface particles of sand were ejected, leaving a crater which steadily enlarged. Some of the ejecta from the crater formed a lip and some particles travelled further to land on the adjacent sand surface. More material accumulated on the downslope side of the crater to form an asymmetrical lip which, after 52 drops, had the characteristic shown in Figure 5.23. After a time, the crater did not appreciably increase in size, most of the material dislodged by a drop falling back into the crater.

The final crater was fairly symmetrical in the vertical section but had an asymmetrical lip. The dimensions were: diameter at neck 5.0 mm; diameter at bottom 3.5 mm; depth 9.3 mm (excluding lip); diameter of lip 19.0 mm (Figure 5.24). The quantity of rain falling was 330 drops each of 0.048 g = 15.84 g, and this dislodged 0.132 cm^3 of wet sand = 0.205 g dry sand. Infiltrating water seeped through the block and emerged downslope.

Figure 5.23. Sand block from Experiment 4, showing crater formed with asymmetric lip.

Conclusion: Under the control conditions used, raindrops falling upon a sloping sand surface can produce a relatively deep crater and deposit more ejecta to the downslope than the upslope side. This indicates the erosive power of rainfall and demonstrates a mechanism by which material may move downslope as part of the mass-wasting process. Since, after a time, the crater ceased to grow appreciably, it is apparent that the process will be limited except when crater form is disorganized by forces other than those of the drops falling directly onto it, which in the field would most usually be drops falling adjacent to the crater.
Systems interpretation: The initial surface is modified at the moment of impact of the first drop, and the crater formed, together with ejecta, becomes part of the morphological subsystem for the remainder of the experiment. The cascading subsystem is the mass and energy of water striking the soil block at the crater, to be re-distributed over the surface of the soil or infiltrating into it.

The stabilization of the crater during the run indicates an equilibrium condition in which mass and energy input are absorbed within the crater, so that the subsystem does not continue to change form. This process has a relaxation time of less than 25 minutes. The effect of continued running of the experiment would cause such changes in the subsurface system that eventually the surface, together with the crater, would be distorted from beneath.

As in Experiment 3, optimum potential energy exists as ejected particles reach the apex of their trajectory, with particles ejected in a downslope direction having greater potential energy than those ejected upslope (Figure 5.25).

Figure 5.24. Section of sand block from Experiment 4 showing shape of final crater.

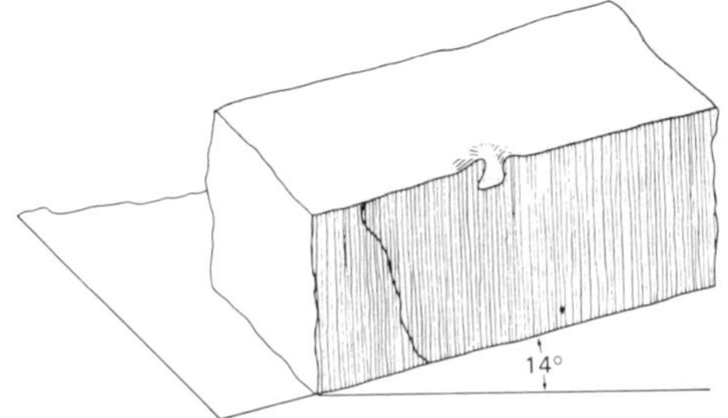

In both the Experiments 3 and 4 it is possible, therefore, to think of the crater itself as the system. In Experiment 3 the crater is a system only momentarily and requires careful consideration. In Experiment 4 the crater, like the soil surface as a whole, is related to a negative-feedback mechanism in which the form achieves stability and can exist in equilibrium with the continued input of mass and energy.

Figure 5.25. Canonical diagram for crater system on a sloping, planar surface (Experiment 4).

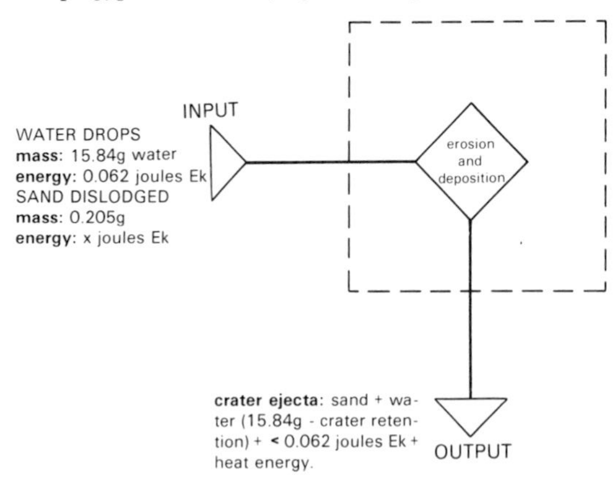

5.4.4 Sheetwash and effect of vegetation on runoff. It seems doubtful whether sheetwash occurs in nature in the form of a uniform sheet of water moving particles downslope. The effect of sheetflow can be studied in the rainfall simulator using a sloping, plane surface, and it is instructive to compare runoff on a bare, vegetation-free surface, with that on a vegetated surface using otherwise identical controls.

Experiment 5

Purpose: To demonstrate the nature of runoff and erosion on unvegetated and vegetated surfaces.
Apparatus: Standard laboratory sand (medium); a rainfall simulator to give a uniform coverage of fine precipitation; a base-board with gable-end pieces allowing a wedge-shaped watershed to be formed, as in Figure 5.26; balance; drying oven; glassware.

Figure 5.26. Condition of unvegetated surface at end of experiment (minute 16), with less eroded vegetated surface to the left.

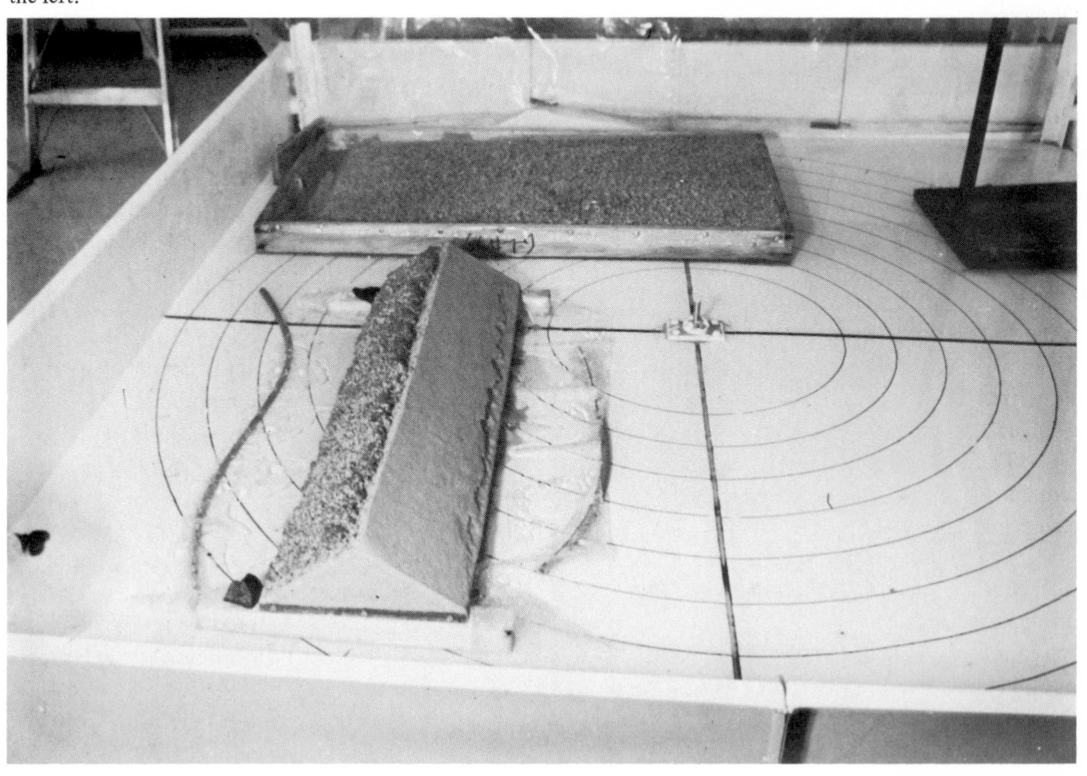

Procedure: The sand is dampened and consolidated on the base-board and accurately shaped into a water-shed form by smoothing with a metre-stick resting on the eaves. One slope is sown with mustard and cress about four days before the experiment; the other slope is kept free. This may be done while the model is in the simulator trough, and it may be kept damp by sweeping the rain-delivery beam across it momentarily twice a day, care being taken not to cause erosion.

The model base should be stood on a plastic apron to facilitate collection of eroded material, and some barrier to debris movement, such as rubber tubing, should prevent it from washing off the apron. After the run, measurements of erosional features are taken and deposited material is dried and weighed.

Controls: Medium sand, modal at 0.25 mm (see Appendix); bed side 100 cm long by 10.5 cm high to ridge, giving 18 cm long slope facets with an angle of 35° (=70% slope; sin angle = 0.583); precipitation drop size modal at 0.8 mm; drop-fall 196 cm; intensity 40 mm/hr; duration 16 minutes.

Observations and measurements: These are as shown in Table 5.4. Post-run collection, drying and weighing of eroded material gave 157 g for the unvegetated slope and 97 g for the vegetated (Figure 5.26).

Conclusion: These two slopes represent in some degree the two concepts of overland flow as put forward by Horton (unvegetated) and Kirkby (vegetated). The Horton model required about five minutes for infiltration to build up the ground-water body to a point at which runoff could occur. This involved a narrow zone near the base of the slope that momentarily produced rills, which then changed to a general basal sapping process.

The zone of no erosion, called X_c by Horton, remained unchanged during the experiment, excepting loss of width by erosion.

The vegetated surface showed first signs of surface activity at about the same time as the unvegetated surface. Runoff was more concentrated into flow lines, giving a sub-uniform spacing of 3-4 cm between rills. Although these broadened into scallop-like forms they remained more or less individually distinct.

Table 5.4. *Effect of sheetwash on unvegetated and vegetated slopes*

Minute	Unvegetated slope
0–5	No sign of surface water on slope (except wetting).
5–6	Storage water appeared towards bottom of slope and within a few seconds rill development and basal seepage began, followed by basal sapping. Meanwhile upper slope (zone of no erosion, X_c) remained only damp.
6.5–16	Erosion zone slowly extended upslope, with rills giving way to a continuous zone of sapping to reach a maximum uphill extension of 3.0 cm.

Minute	Vegetated slope
0–6	No sign of surface water.
6–7	Small rills appeared at base of slope. These deepened and widened to form scallops 2 cm in width and 3–4 cm apart.
12	Vegetation mat plus root soil slid *en masse* downslope a short distance.
7–16	Scallops had worked back 1–2 cm.

The sudden slipping of the surface in the central part of the face exposed a band of vegetation-free sand along the model crest, and the rills at the edge renewed their activity for a short while. This slipping of vegetation and root soil is probably due to a process resembling soil-creep and suggests that a lubricated layer occurred between root mass and underlying groundwater table. It seems likely that flow was occurring in the root zone, or just beneath it, and erosion could most readily occur in lines of concentration.

This may, therefore, be taken as a simple type of throughflow as required by Kirkby's model. The difference in values of eroded material, in which the vegetated surface yielded only 62% of the unvegetated surface, reflects differences in types of surface runoff process and incidentally underlines the protective function of vegetation.

Systems interpretation: The use of a single test-block for two different test surfaces effectively produces two models. A sharp water-shed separates surface conditions but the groundwater body will be common to both surface facets. Some distortion of the water table can be expected due to different infiltration and seepage characteristics on either side of the soil block.

The situation may be represented as in Figure 5.27, in which the slope facets are separated, the water table is regarded as common to both, and the erosion zones are separated. It may be preferred to modify the systems model to separate groundwater into two parts.

In the Horton model the precipitation input (1557 g) is seen to infiltrate or run off the surface. Some of the surface flow will produce initial rills. Water that infiltrates on the Horton model surface will move via the groundwater body to the zone of effluent seepage, which is coincident with the surface runoff zone, and basal sapping, with components of slump and mud-flow, will rapidly mask and replace the rilling process. The process will slow down due to a negative-feedback loop, which will initially be involved with the destruction of rill forms, and then with the removal of material generally from the edge of the slope until groundwater seepage is occurring from a stabilized slope. This incorporates the steep backing slope and a gentler footslope, the latter of which is coincident with the water table and base-level of erosion (as determined by the edge of the model baseboard). There is a relaxation time of less than 16 minutes, during which 157 g of sand have eroded.

In the Kirkby model precipitation will infiltrate or move downslope amongst the roots of the vegetation. This is analogous to throughflow in a deep field soil. Infiltrating water will join groundwater and seep out at the foot of the slope, but will not completely destroy the concentrated flowforms associated with water movement by throughflow. This system did not quite reach equilibrium conditions and relaxation time is therefore greater than 16 minutes. Sand produced as part of the output is 97 g in 16 minutes. A negative-feedback loop may be deduced to exist in which stable forms will eventually develop, possibly as broad scallops backing a footslope coincident with the water table.

Evaporation in 16 minutes is negligible, so water output will be equivalent to water input to the total test-block, minus water held in storage in the groundwater body.

The energy input of about 60 J will be used in transporting particles and overcoming friction, and the balance will be dissipated as heat.

An alternative model (Horton or Kirkby) may be employed, using a larger sand tray* with a surrounding wall (122 cm x 63.5 cm x 10.0 cm). This is filled with a wedge of sand, leaving a few centimetres at the foot for deposition of eroded material (Figure 5.28).

The sketch of surficial features of the Horton model was taken from a student's notes and the observations accompanying it were:

> In the above experiment rainfall at a rate of 0.026 inches/minute (40 mm/hr) was administered ... After about 1.5–2.0 minutes it could be seen that the water table had reached the surface everywhere below a line (at) approximately 35 cm downslope. Above this line the infiltration rate was greater than the rainfall rate so there was no surface storage or erosion (zone X_c). After 3.5 minutes of rainfall, surface storage had built up below this X_c zone and had begun to lead into overland or sheetflow around 75 cm downslope. At the bottom of the slope, concentration of water had begun to form rills. The depth of the flow was *c.* 1 mm, with a velocity of 127.5 cm/minute. It was of a turbulent nature at the beginning and changing to laminar flow further downslope ...
>
> The sequence was, going from top to bottom

Process:	infiltration →	surface storage →	sheetflow	→ concentrated flow (erosion) →		deposition
Form:	X_c	→ puddles, ponds →	standing waves →	rills →	small channels	→ deltaic material ...

* This is constructed from plywood lined with polythene, and may double as a stream-table.

Figure 5.27. Systems interpretation for Experiment 5.

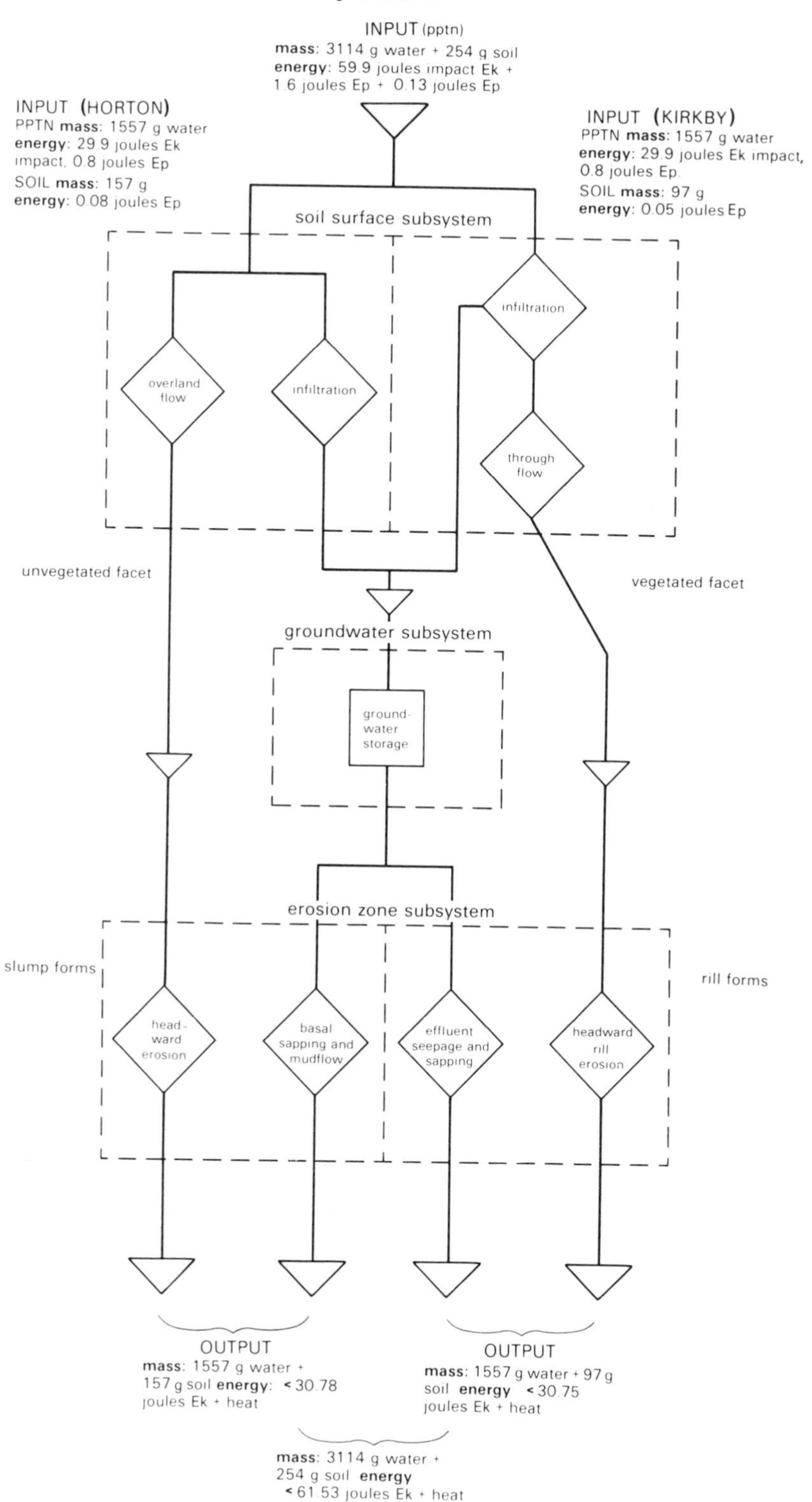

5.4.5 Infiltration, eluviation and illuviation. This experiment makes use of natural soil removed with care from a field location and placed with as little disturbance as possible into a specially made tray.

Experiment 6

Purpose: To demonstrate the process of infiltration in a natural soil.

Apparatus: Soil sample; rainfall simulator to produce a uniform, high-intensity precipitation; a soil tray 20 cm x 20 cm x 10 cm with an outlet pipe from a wall near the base, the outlet pipe to curve downward in order to feed into a catch-bottle, the opening of the bottle shielded from precipitation by a cowling attached to the pipe (Figure 5.29); a raingauge; an accurate balance; various glassware.

Procedure: Moisture content of soil sample was determined and the bulk of the soil was placed in the tray, leaving a lip of about 1 cm around the sides. A catch-bottle was placed beneath the cowl

Figure 5.28. A student's derivation of the Horton model.

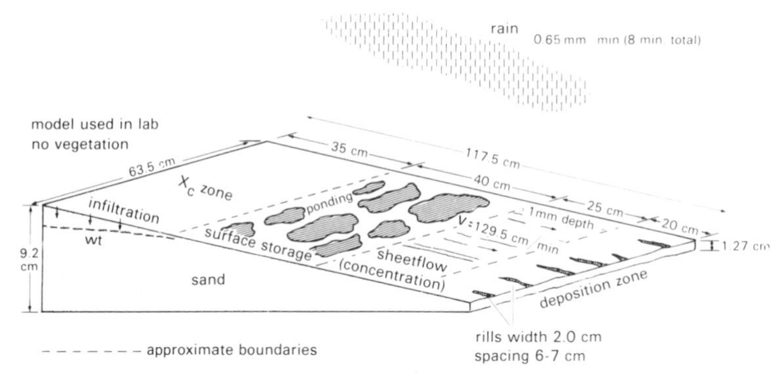

Figure 5.29. Apparatus for infiltration study (Experiment 6).

with the end of the outlet pipe centred over the bottle neck. Rainfall characteristics were pre-set, but the precipitation was monitored with a rain-gauge during the run. The run was commenced and note was taken of development of phenomena. After the run the catch was analysed for water and solid components by volumetric methods, followed by oven-drying dessication and gravimetric analysis.
Controls: Precipitation intensity 0.135″ in 7.075 minutes (29 mm/hr); drop size modal at 1.0 mm; distance of drop-fall 195 cm; duration 7.075 minutes; field soil.
Observations and measurements; The pre-run moisture content of the soil was 31% and this built up to field capacity in about 1 minute after the commencement of the run. By 1.5 minutes the first drops of water entered the catch-bottle and at 3.5 minutes surface storage commenced on the soil. By switch-off time, ponding at the surface was extensive. The total catch for 7.075 minutes was 5.26 g, of which 0.075 g were solids.
Conclusion: A definition of *infiltration* must be used to quantify conclusions. If infiltration is taken as being the passage of precipitation water through the surface of the soil, then 136 g of water infiltrated in 7.075 minutes = 19.2 g/minute, over a 400 cm^2 surface.

If infiltration is taken to include the movement of water downward through the soil, then 136 g moved downwards through 9.0 cm of soil, some reaching the soil base in 1.5 minutes, and 5.185 g ran from the base of the soil block in 7.075 minutes. This gives a time-averaged infiltration rate of 0.73 g/minute through a 400 cm^2 section.

As soil moisture-content increased to a stage at which surface storage commenced, it is clear that precipitation intensity was greater than rate of infiltration through the soil mass.

The experiment demonstrates infiltration, and analysis of the data illustrates the complexity of the infiltration process and the difficulty of measuring its component parts.

During the experimental period of 7.075 minutes, 0.075 g of fines were washed out of the sample into the catch-bottle, representing part of the eluviated soil material, note being taken of the likelihood of fines being trapped in the base of the soil mass during lateral movement towards the outlet pipe.
Systems interpretation: The morphological subsystem is taken as being the total soil block of 400 cm^2 area and 9 cm depth, together with the pore-spaces in the soil. The cascading subsystem input is precipitation mass and kinetic energy of impact (136 g and 2.62 J), plus immediate potential energy of this water at the soil surface (0.12 J), together with soil dislodged at various depths in the soil (mean depth of soil, $\bar{H} = 9/2 = 4.5$ cm) assessed from the catch as 0.075 g (but it would be higher since some material would merely move within the system), together with its potential energy of 3.31×10^{-5} J.

Output is 5.185 g water plus 0.075 g solids with less than $2.74 + 3.31 \times 10^{-5}$ J kinetic energy, plus heat energy (Figure 5.30).

Eluviation of solids may enlarge some pore-spaces and illuviation may block others. The changes in the morphological system are therefore of two kinds, but the net result will be diminution of infiltration rate to an equilibrium condition, indicating negative feedback. It is not likely that equilibrium was reached during the experiment.

Figure 5.30. Infiltration and illuviation system for Experiment 6.

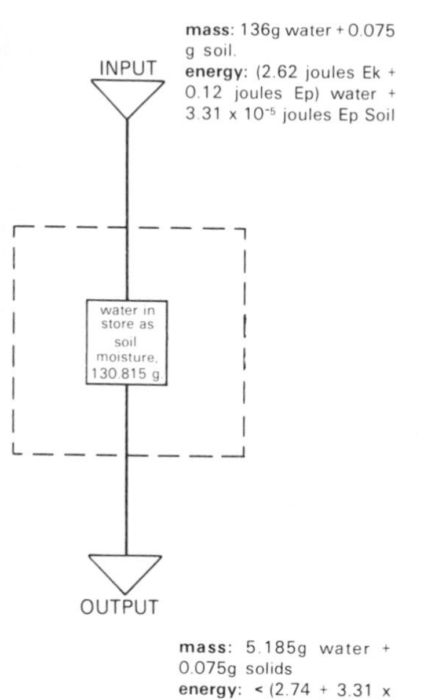

6 The rainfall simulator in analogue modelling

6.1 Rainfall-dependent analogue forms and processes (introduced as category (2) in section 5.4)

The material of the model is usually a medium sand which is an analogue of a natural rock or sediment, and two basic form-types can be developed in the rainfall simulator, the stream or fluvial feature, and the slope or interfluve feature. In these models scalar properties become very important since processes and forms may not jointly equate with reality.

The linear dimensions of medium sand grains will possibly average about 0.25 mm. If a feature of 1 m width represented a piece of territory of 10 km in width the landform scale is 1:10 000, and at that scale the sand grains would represent boulders of 2.5 m across. If a *river* were to be developed at this scale it would be less than 10 km in length, eroded in a bed of 2.5 m boulders.

This scalar difference in geometric properties is common in analogue models used for the study of reality and is acceptable providing processes are real. In a small, 1 m model, water flow is likely to be in the laminar mode, and this cannot be related to grain size. Similar considerations apply to other aspects of the models and a matter of semantics arises. For example, if the model exhibits a *cliff* – what can the term *cliff* acceptably be used for? In reality, very large precipice faces might exceed cliff size, and small vertical terrace rises of a metre or so are too small to be so named. An understanding of size-limitations for earthform names is common, and arises from the fact that many such form names are derived from vernacular usage and have never been precisely defined with a view to inclusion of form irrespective of size. In physics and chemistry such words exist but also have usually been precisely defined for scientific purposes, and in the biological sciences such vernacular terms have been replaced by the Latin or Greek equivalent with precise definition. Geomorphic terms suffer from lack of precision and this is particularly evident when scale changes are involved.

A cliff, according to the *Oxford English Dictionary*, is a steep rock face, usually facing the sea. Here, in a general sense, no size is stipulated, but the process of erosion by sea-action is implied. Cliff is not defined in the *Encyclopedia of Geomorphology*, but is linked to marine cliff – which is not given an individual entry. It is certain, however, that vertical faces in a 1 m dynamic model are not cliffs in the general or technical sense, except as analogues.

Processes in earth science do not suffer from the same problem. They have probably not been represented by a vernacular term, and have usually been properly defined or explained. Even so, in dynamic models of scaled-down type, they should always be questioned. For instance, the formation of a delta in the model appears to observe the essential aspects of delta formation on a larger scale in nature. Material is carried out from the coastline by a current of water and settles at some distant point depending upon its mass, volume and density. In such a case, *delta* may be an incorrect term to use for a 5 cm long feature, but in so far as processes of formation are concerned, they agree with the processes at work in the prototype.

A process may accurately duplicate a natural large-scale process and yet produce a result that cannot be given the concomitant landform name. For example, material in the model may slump out to produce a vertical face. The slumping process can be a true slumping process, while the vertical face is too small to be called a cliff.

There are thus several conditions in scale models used for demonstration of natural features:

(i) in which form resembles a real form but is too small to be accepted as such except by analogy (cliff, river),

(ii) in which a form can be accepted as a real form, there being no understood size limit (terrace, rill),
(iii) in which a process is exactly the same as its real equivalent (delta formation, bank undercutting),
(iv) in which a process does not represent a real equivalent (bank fluting due to surface tension).

Of these, case (i) is most obvious and many examples exist, (ii) is not so common, (iii) is frequent and (iv) is very rare. The four conditions can all exist in a given model, and can lead to conceptual confusion for more analytical observers.

Types of systems or properties (usually scaled analogues) that can be represented in the simulator are rill, gully and river formation, rill erosion, drainage-net development, centripetal drainage, radial drainage, badland formation, erosion of domes, anticlines, synclines, cliff formation, waterfall formation, headward erosion, peneplanation, equilibrium, base-level control, scree formation, landslides, mud-flow, allometric growth, thresholds, storage, water-table effects, terrace formation and a range of effects demonstrating the erosion of multi-stratified rocks.

Some of the above are more readily produced in a flume, others are essentially simulator models. Some of the most characteristic are badland topography, in which an almost featureless landmass is eroded into forms typical of badlands. This is a micro-landform assemblage that runs the gamut of reality and analogue within a given experiment to the extent that a special study can be made of it (for example see Chapter 12 where such forms are studied in quarry sludge).

In the simulator the sand can be conveniently shaped into an island standing in about 3.0 cm of *sea-water*. Attention should be paid to the types of feature that can develop and the initial surface should be shaped accordingly.

One of the main points is that a fairly heavy rainfall-intensity will be required (30 mm/hr or more) so that the water table can rise. This gives a local base-level of erosion higher than sea-level; it permits rills and gullies to develop and forms generally to adjust to steep slopes of parallel retreat and gentler bajadas.

6.1.1 Rill and gully formation. Sheetwash and rill forms may be regarded as real micro-features such as may be found in patches of fine sediment at the foot of new roadside cuttings, around the edges of ponds and in fine beach material. If, in a simulator model, these are looked upon as representative of larger-scale reality, the type of drainage feature that develops in semi-arid areas, then the model is scaled, and if materials or processes differ from the prototype, it is an analogue.

Essentially, the model will resemble Figure 6.1 and attention must be given to interpretation.

Experiment 1

Purpose: To study by means of a dynamic model the development of rills, gullies and related phenomena in a semi-arid landscape, firstly with constant sea-level (base-level of erosion) followed by falling sea-level.

Apparatus: Rainfall simulator to produce steady, fine, high-intensity rain; trough to contain sea; medium sand; raingauge; azimuth bar; lab timer; probe; various glassware.

Figure 6.1. Plan of island used for Experiment 1, surveyed with azimuth bar and probe.

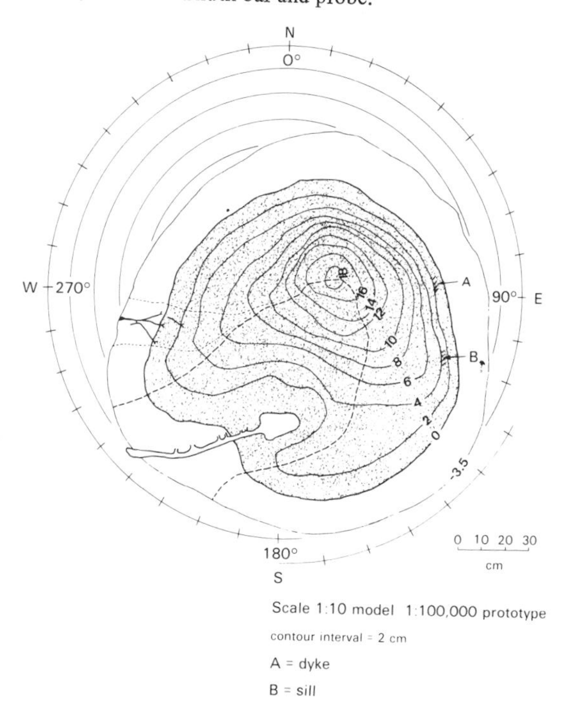

Procedure: The rainfall-delivery system was pre-set and held in readiness; sand was placed in the simulator trough in such a position that sea could exist all around it, but any part of it could be reached for measurement. The sand was shaped into an island form, account being taken of (*a*) processes and forms to be developed, (*b*) the relationship of groundwater table (GWT), infiltration characteristics and potential surface-flow zones. Sea-level was built up, the coasts were trimmed and the topography was measured by azimuth bar using sea-level as datum and plotting horizontal distances from the azimuth bar centre on a radial grid (Figure 6.1). The precipitation was monitored by a rain-gauge during the run.

The initial island form was shaped to include: some steep coastal zone; some flatter coastal zone; an inland basin low enough to enable groundwater to produce a lake; a mountain high enough above water table to permit rapid and continued infiltration; some offshore shelf; a bay. In addition, strata of resistant rock were represented by molded blocks set into steep-sloping and gently-sloping parts of the model (Figures 6.1 and 6.2). An apron was placed in the bay to collect deltaic material.

At the conclusion of the run with constant sea-level, measurements were taken of the forms that had developed and the experiment was then continued with falling sea-level. At the conclusion of this second phase, final measurements were taken of relevant erosional and depositional forms.
Controls (phase 1 constant sea-level): Horizontal scale 1:10 000; vertical scale 1:10 000. Island mass: medium sand (analogue for solid rocks), see Appendix, Figure A.1; dimensions *c.* 130 cm diameter (horizontal scale 1:10 000 giving a prototype island of 13 km); area 12 240 cm^2 giving a prototype area of 150 km^2; relief 18.5 cm (vertical scale 1:10 000 giving a prototype height of 1.85 km or *c*. 6000 ft); lake hollow 0.1 cm above sea-level (ASL) (=10 m ASL in prototype); ridge between lake hollow and sea at bay 0.6 cm (=60 m ASL in prototype); offshore slope greatest width 28.6 cm (=2.86 km in prototype), mean width

Figure 6.2. Initial island form for Experiment 1.

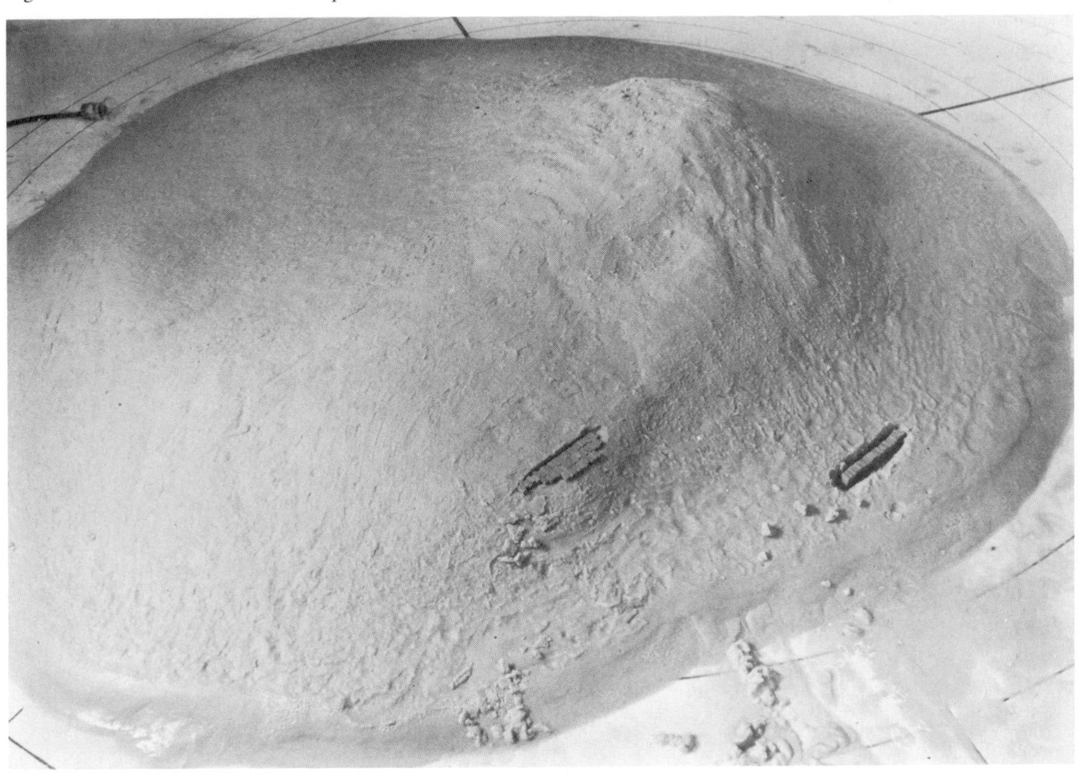

16 cm (=1.6 km in prototype); ratio of slopes model/prototype = 1:1; sea-depth 4.0 cm (=0.4 km in prototype); precipitation drop size median 0.2 mm diameter (finer than natural rain but of no scalar significance); precipitation intensity 37.5 mm/hr (natural intensity); drop-fall distance *c*. 2.0 m (=20 000 m in prototype, but scale only of significance if used to calculate terminal velocity of drops); duration of constant sea-level run 38 minutes; water table before run 0.0 cm at coast and 0.1 cm at lake hollow due to capillarity and unlikely to be greater elsewhere in the model. (It may be assumed before the run that sheetflow over the model will be laminar, and that rill-flow will be laminar or laminar/turbulent and capable of entraining sediment. The relationship between gravity and particle movement and the factors of resistance, capillarity effects, surface tension, trajectories and other dynamic and kinematic attributes can be assessed by applying the general considerations discussed in Chapter 3. The question of iconic or analogue processes and forms can be dealt with during *observation* and *conclusion*.)

Observations and measurements (phase 1 constant sea-level): Immediately after commencement of precipitation, infiltration occurred all over the island, and the glisten of surface saturation showed first around coast and lake hollow. At MINUTE 3, the first rills appeared on the northern shore of the lake (the lake being formed indicating rise of the water table). The water table was about one-third of the distance between the sea and the peak of the mountain on the northern flank, showing GW rise of 6 cm. By MINUTE 4.5, the lake and sea were separated by a ridge of 7.6 cm width as the lake expanded. By MINUTE 12.5, strong flowlines were developing around the eastern edge of the lake. By MINUTE 15, slumping was occurring to the east of the island cutting back to a resistant sill (Figure 6.3); the lake was flowing out to sea over the ridge, leaving a spatulate tongue of sediment. Flowpaths occurred around most of the lake. There was rilling on the shore slopes to the north of the island. By MINUTE 30 a clearly defined rill system had developed but without depth of incision. The lake and other systems, except slumping, appeared to

Figure 6.3. Experiment 1 at minute 15 of phase 1.

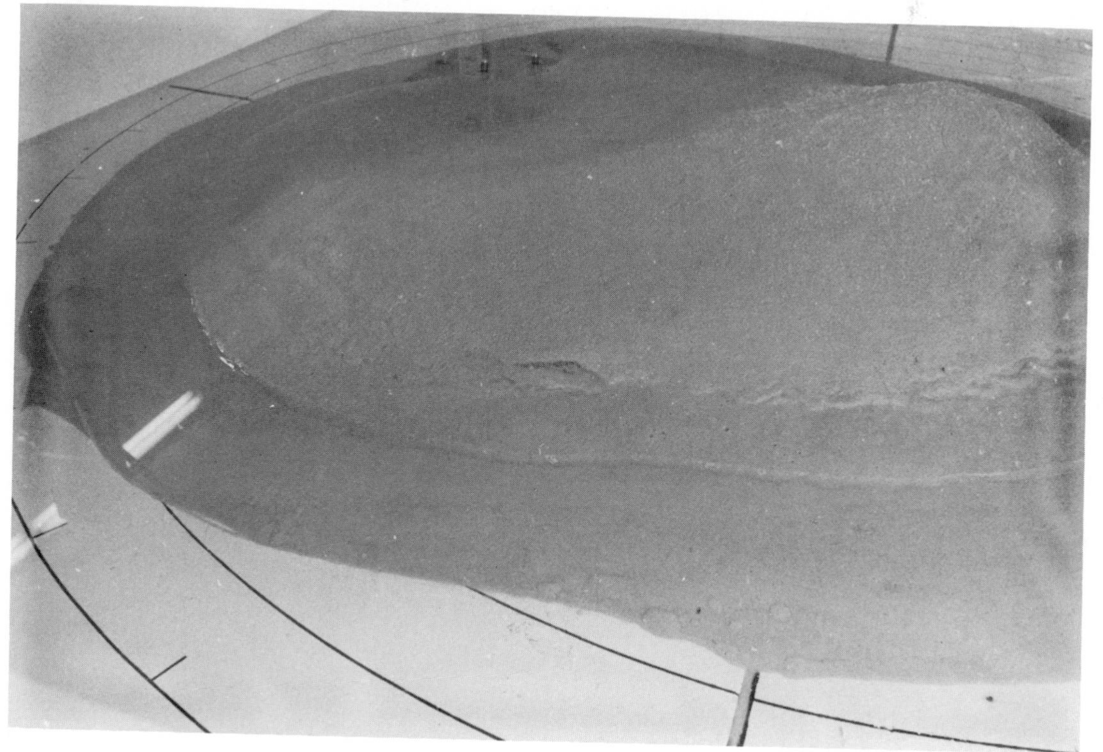

be in a condition of stable equilibrium at this time.

In general the longest rills developed around the lake, where they were about 25 cm in length. The rill system was otherwise coastal. The lake reached an optimum size of 25 cm × 25 cm once it began to flow out to the sea, but some infilling of the lake from the centripetal rill system occurred (Figure 6.4). Slumping occurred along eastern slopes and failure affected resistant rocks beneath the steeper surface, but not the dyke beneath the gentler slope.

No delta reached the outer edge of the offshore slope (Figure 6.5) during this part of the experiment, but a tongue of sediment derived from the lake was deposited in the main river channel, measuring 4 cm × 3 cm, and deltas from rill erosion formed in the lake.

The water table rose a few millimetres above sea-level in the lake area, and about 6 cm ASL in the highland mass.

The drainage system comprised a set of rills flowing down the lower slopes, together with a stream–lake system carrying its own tributaries.

At the end of phase 1 of the experiment (sea-level constant at 0.0 cm), the various parameters were: island area 12 240 cm^2 (unchanged); diameter *c.* 130 cm (unchanged); relief 18.5 cm (unchanged); offshore slope unchanged in plan, but added to by some coastal erosion; number of coastal rills 20 (with a distribution by quadrant of NE 9, SE 0, SW5, NW6); average length of coastal rills 6.5 cm; main stream–lake system: stream 5.0 cm long, 2.5 cm wide, 0.15 cm deep, no tributaries; lake 25 cm E–W, 25 cm N–S, 0.1 cm deep, number of tributary lake rills 9; average length of lake rills 18 cm, width 10 cm, depth 0.1 cm; also 4 leads to west of lake and rills to north and east; mean width of coastal zone of saturation 24 cm; mean width of coastal erosion surface 6.5 cm. Other parameters were the width (9 cm) and vertical extension (1.5 cm) of failure at the slump zone on radius 110° (Figure 6.6).

Figure 6.4. Experiment 1 at end of phase 1, showing centripetal drainage (note differential compaction to left of lake caused by hand pressure).

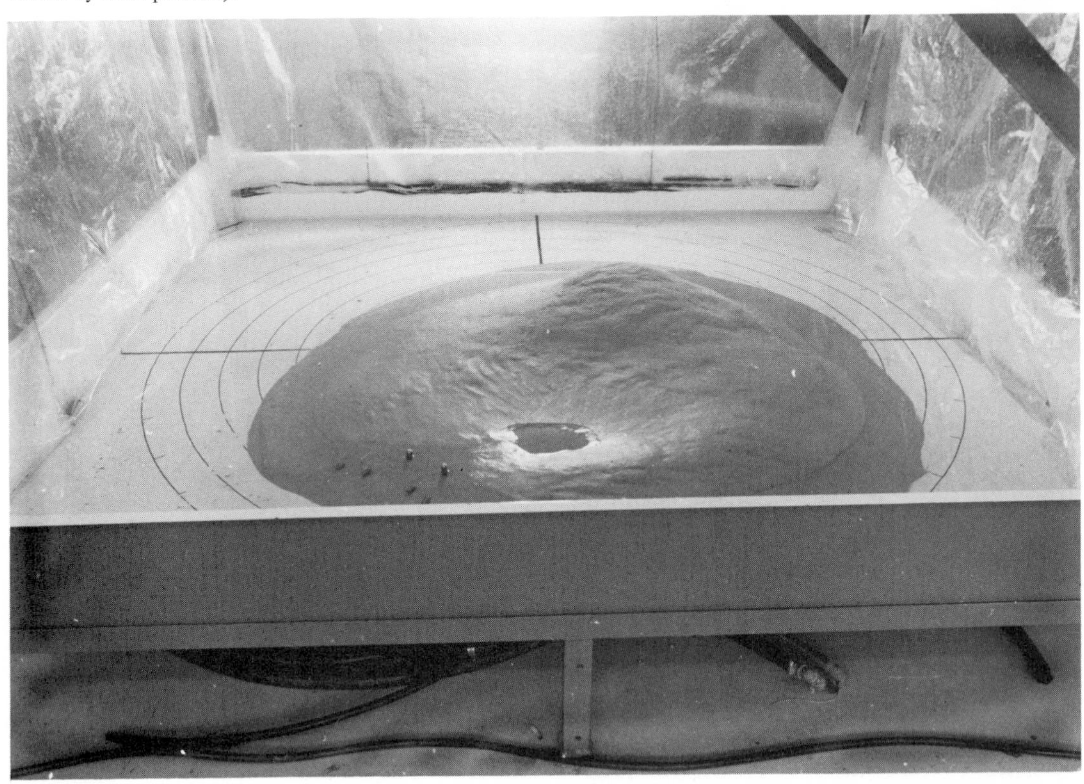

Conclusion (phase 1 constant sea-level): The development of dynamics in the model is dependent upon the state and form of the groundwater body. Since there were no endogenic or temperature-change forces at work the rainfall, infiltration and groundwater movement gave the initial impetus to landform change. As the groundwater table rose within the model several significant processes developed: movement of groundwater around the perimeter of the island caused failure of the coastal cliff and rapid establishment of a new coastal slope by mud-flow or basal seepage. This was a condition that reached equilibrium within a few seconds of commencement of the run. A second process was formation of a lake or surface storage as the GWT intersected the island surface in the hollow zone to the south; a third process was the development of rills wherever surface runoff occurred on the lower slopes due to the raising of the GWT to the surface; as the GWT rose to the level of the hard blocks of dyke on the steeper eastern slopes, seepage at the base of the dyke caused failure and a slumping process, and in the lake system, where groundwater was able to collect by seepage and flow out at a point of concentration, a stream developed which entrained sediments and redeposited them within its own channel. The zones of optimum rill development were on moderate slopes of *c.* 12° in the NW quadrant.

Land above the water table remained unaffected by surface processes, as infiltration produced no discernible changes.

Processes within the model achieved an equilibrium state at varying times. By the end of the run, rill, lake and river development had ceased or were imperceptibly slow. Slumping at the zone of effluent seepage was the only process noticeably continuing until MINUTE 38.

Controls (phase 2 falling sea-level): As for phase 1 excepting sea-level, which was lowered throughout the run from 0.0 cm to −3.5 cm at a drop rate of 2.8 cm/hr, and duration of run 75 minutes.

Observations and measurements (phase 2 falling sea-level): Erosion and deposition recommenced

Figure 6.5. Experiment 1 showing sedimentary deposit in outlet stream channel.

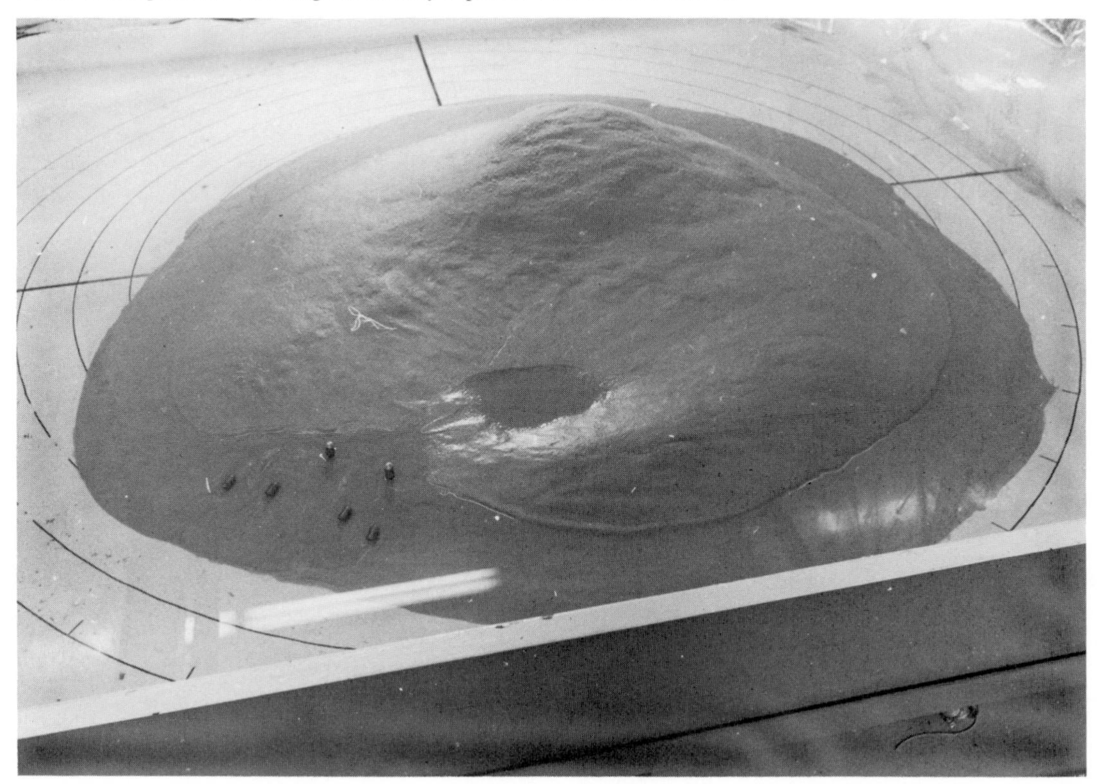

actively throughout the island. At MINUTE 8, the shore slope was extending to the receding sea-level. By MINUTE 13, the rills were incising strongly, and the main delta was growing. By MINUTE 20, the rills to the north were incising deeply and the main stream from the lake was extending towards the sea over the newly exposed land surface (Figure 6.7). By MINUTE 21, the major sea-rill was forming on a radius to south of west. At MINUTE 26, all rills were continuing to grow. By MINUTE 35, the main stream was beginning to drain the lake and tributaries were forming along it. The water table was about 3.6 cm above the 0.0 cm datum. By MINUTE 47, there were very well-developed small rills to the east (Figure 6.8). By MINUTE 52, about six tributaries had developed along the main stream. By MINUTE 60, new networks were developing in the NE quadrant. The main stream was flowing strongly. By MINUTE 69, deltas were developing all around the island. At MINUTE 75, switch-off.

Figure 6.6. Experiment 1. Slumping at the sill.

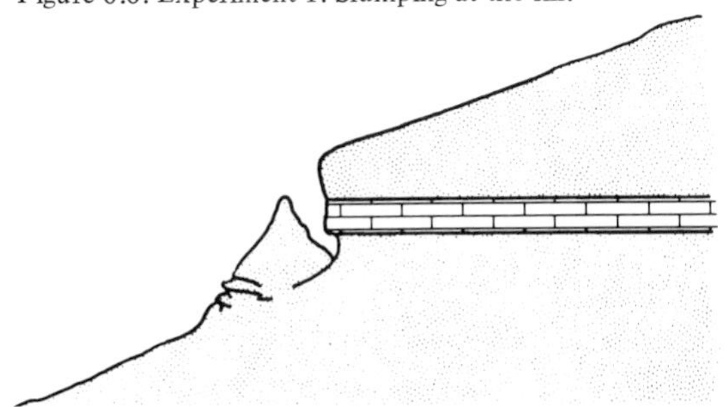

The renewed development of all aspects of erosion and deposition was noticeable as sea-level dropped.

Existent rills became stronger, incised and lengthened and developed their own tributaries. New rills formed all around the island in the zone of saturation and tended to be longer on steeper slopes away from the lake system. The lake shrank and drained off as the GWT fell. Edges closed in owing to sedimentation from surrounding slopes. The river extended in length, remained of more or

Figure 6.7. Experiment 1 at minute 20 of phase 2. Outlet stream from lake extending over newly exposed land surface.

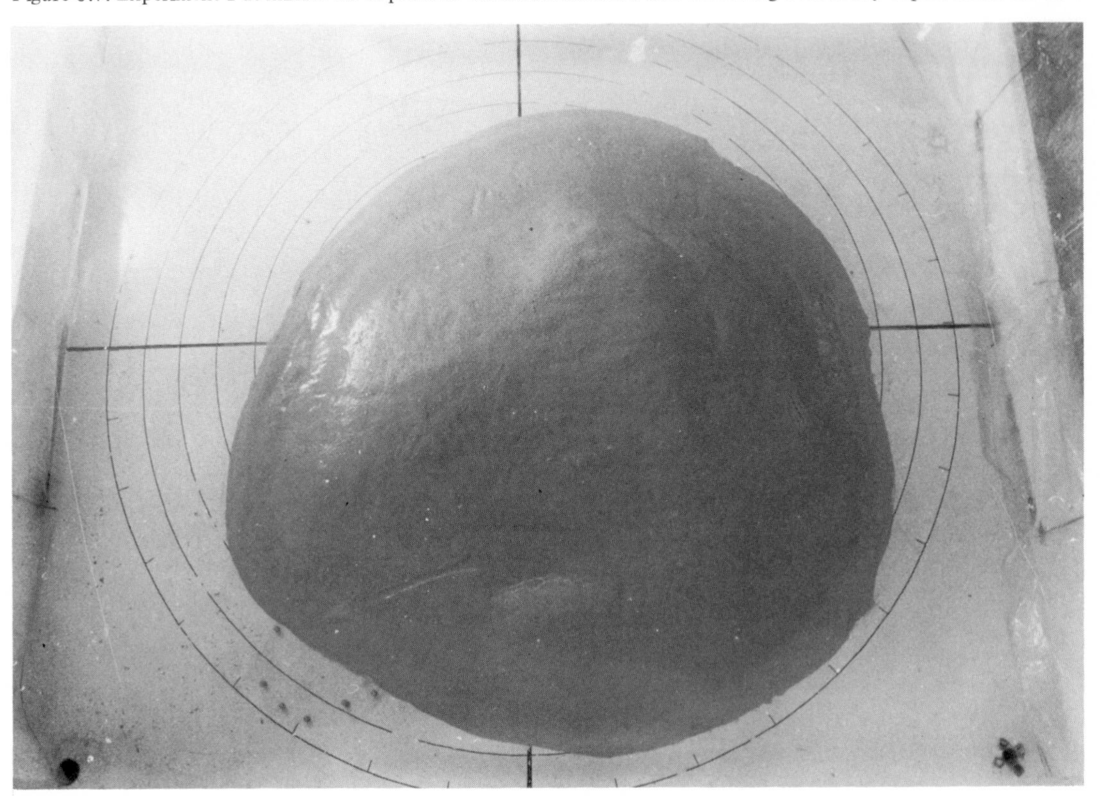

less constant width and developed progressively more tributaries.

The emergent offshore slope was the main surface for rill formation. The small backing cliff became smoothed out to some extent and most rills cut back as far as that faint flexure. Rills existent on old land surfaces were usually a survival from phase 1.

Equilibrium conditions in rill-network development were not existent at the end of the run; the lake continued to drain and fill to switch-off; the water table continued to fall and delta growth continued to the end. The mountain massif diminished in area by attrition of the lower slopes also to the end of the run.

At the end of phase 2 of the experiment (new sea-level at −3.5 cm) the various parameters were (data collected by azimuth bar etc.): sea-bed extension of island negligible; land extension of island resulting from sea-level drop averaged about 18 cm all round, giving an increase in diameter from 130 to 166 cm, with an area increase from 12 240 to 19 900 cm^2; reduction of mountain height 0.5 cm; increase in relief 3.5 cm (sea-level drop) −0.5 cm (mountain reduction) = 3.0 cm giving total relief of 21.5 cm; eroded material, based on deposition beneath original sea-level: wet, loosely-compacted material volume* 1456 cm^3 for coast excluding main-stream delta, and wet, loosely-compacted material volume 21.5 cm^3 for main-stream delta = total of about 1477 cm^3 wet, loosely-compacted material = total of about 1285 cm^3 wet, compact material = total of about 1992 g dry sand†; number of clearly defined rills 74, distributed by quadrant SE 27, SW 17, NW 14, NE 16; common rill parameters, spacing 5–7 cm, width of mouth 3–4 cm,

* This is estimated by difference in section between pre-run and post-run undersea surface.

† Medium sand in use yields *c.* 1.55 g dry sand per cm^3 wet, loosely-compacted sand. (1 cm^3 wet, loosely-compacted sand yields 0.87 cm^3 wet, compact sand.)

Figure 6.8. Experiment 1 at minute 47 of phase 2, showing rill development in eastern sector.

length 7–11 cm; some larger rills in NE quadrant are 15.0 cm long; rill at NE diagonal is 20 cm long; rill 10° W of N is 30 cm along its thalweg; rill depth is as much as 5.0 mm at shoulder; main stream 55 cm long, 2.6 cm wide across top of valley; channel 1.5 cm wide, 0.5 cm deep (Figure 6.9); the main stream has parallel right-bank tributaries 2–5 cm long, 1.0–2.5 cm wide (velocity of main stream at MINUTE 72 *c.* 2.0 cm s^{-1}); delta 10.0 cm wide, 2.5 cm seaward extent; slump region at dyke to east: total width of all traces of rock failure 17 cm; major slump material 13.5 cm^3 wet volume; exposed face in front of dyke blocks 6.5 cm wide, 2.0 cm high cliff.
Conclusion (phase 2 falling sea-level): Lowering of sea-level caused lowering of the base-level of erosion, governed at the island's perimeter by sea-level and inland by the groundwater table. A concomitant effect was increase of potential energy available to the island due to a 3.0 cm increase in relief (note that the island lost 0.5 cm in absolute height, possibly by continued settling). It may be deduced that there would be an initial increase in hydraulic head in the groundwater body and, by Darcy's law,* an increase in rate of flow by effluent seepage until the GWT had re-established an equilibrium state.

In this phase, as sea-level dropped continuously until switch-off, it may be assumed that the

Figure 6.9. Experiment 1 at end of phase 2: section across outlet stream.

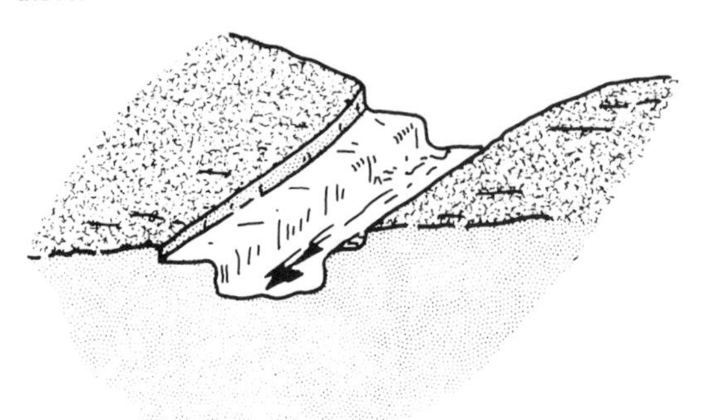

* Darcy's law states that $Q = KA(H/L)$, where Q = discharge or recharge, K = coefficient of permeability (about 1.0 in the sand in use), H = hydraulic head, L = distance from GW crest to lowest point (or point in question) and A = area of unit cross-section.

GWT did not attain an equilibrium state, and that seepage was always greater than during phase 1.

As a result of increased flow from within the island and the falling of the water table from below the old rill-beds, re-incision (rejuvenation) occurred accompanied by some minor headward extension. In general, rills do not cut headward into the zone of no-erosion, X_c,[1] although basal sapping and head-wall failure may produce a small headward extension. The increase in length of rill systems was mostly downwards as rills kept pace with the receding shoreline.

The rill channels were usually about 1 mm deep, except inland where they cut into the old land surface keeping pace with lowering of the groundwater table, and 0.5 mm depths occurred over the inflexion of the old shore-edge. As the rills eroded the newly exposed offshore slope, they were flowing in the interface zone between groundwater and land surface, and were unable to incise deeply. The rills formed a consequent radial drainage pattern around the island giving a drainage density of approximately 0.8 cm/cm^2 for the island, excluding the main-stream catchment. A well-developed dendritic system occurred on radius 260° (Figures 6.10 and 6.11) draining an area of 800 cm^2 adjacent to the perimeter of the main-stream system. The trunk rill is third-order with 9 fingertip tributaries and a network bifurcation ratio of 3.

The rill systems (which on a scale of 1:10 000 would be gullies in the prototype) are fed mainly from the groundwater body and extend by sapping at the sides and head, while running water carries off dislodged material. Walls of the rills are characteristically vertical due to bank failure and to lack of mass-wasting on interfluves, and floors are flat due to control by the groundwater table.

The rill load was deposited mainly on the offshore slope, building out a delta in such a way that the overall diameter of the island at the sea-bed was scarcely increased (see Figures 6.12 and 6.13). The quantity of material thus deposited around the island, including the main-stream delta, was about 1992*g*.

To the south of the island the stream–lake drainage system also became strongly activated by the lowering of the sea-level. The water-level in the lake fell and infilling by material brought into the lake by its own tributary rills shrank the lake diameter. Water was able to concentrate in the main stream from the total catchment of 5240 cm^2 and the stream-channel grew to a length of 55 cm southwards across the old bay bed. It also incised along much of its length, producing a new channel of 1.5 cm width within an old channel (valley?) of 2.6 cm width, becoming a misfit stream. The water surface in this stream functioned as a local base-level of erosion and tributary rills developed particularly along its right bank, receiving seepage

Figure 6.10. Dendritic system formed by end of phase 2 of Experiment 1.

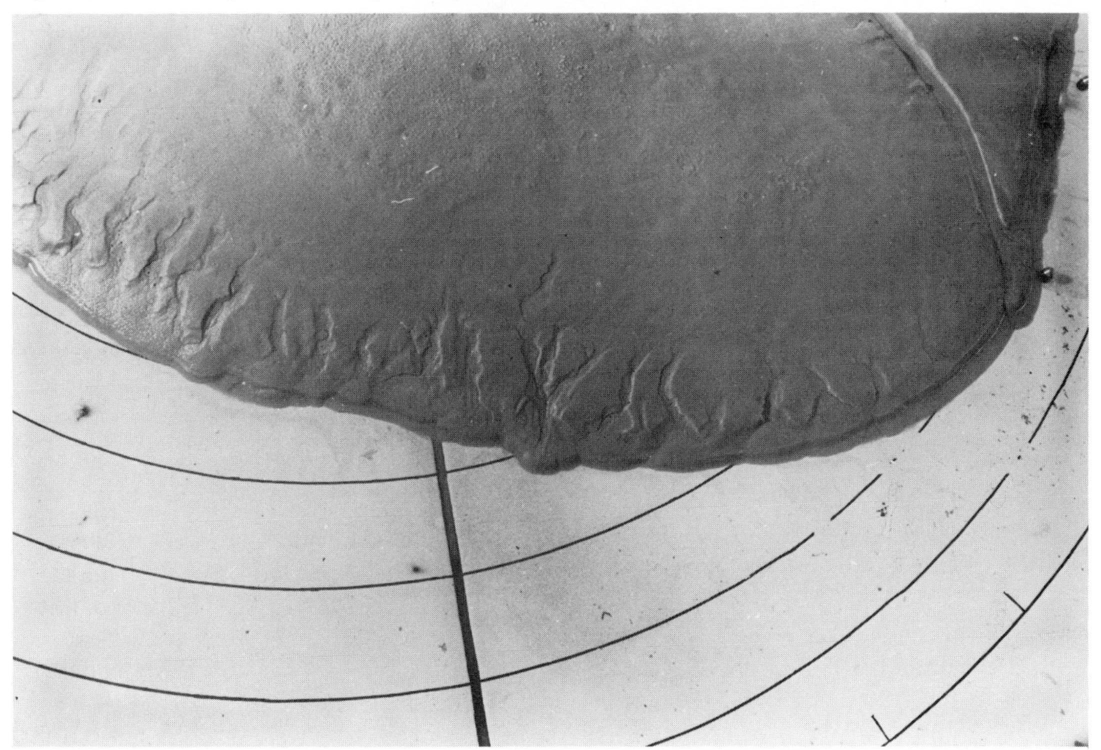

Figure 6.11. The small catchment and drainage net shown in Figure 6.10.

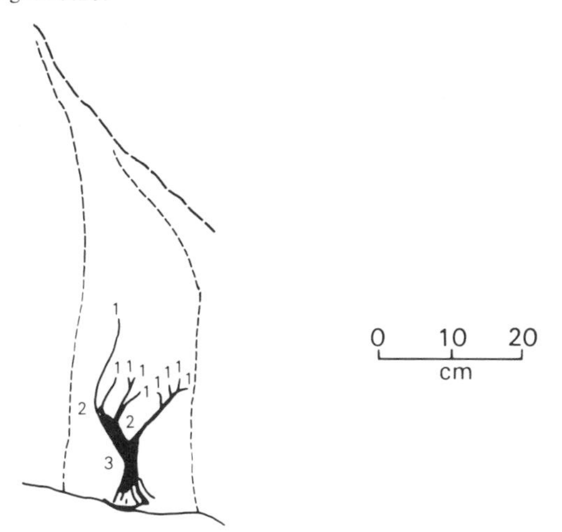

Figure 6.12. Perimeter of island at commencement of Experiment 1.

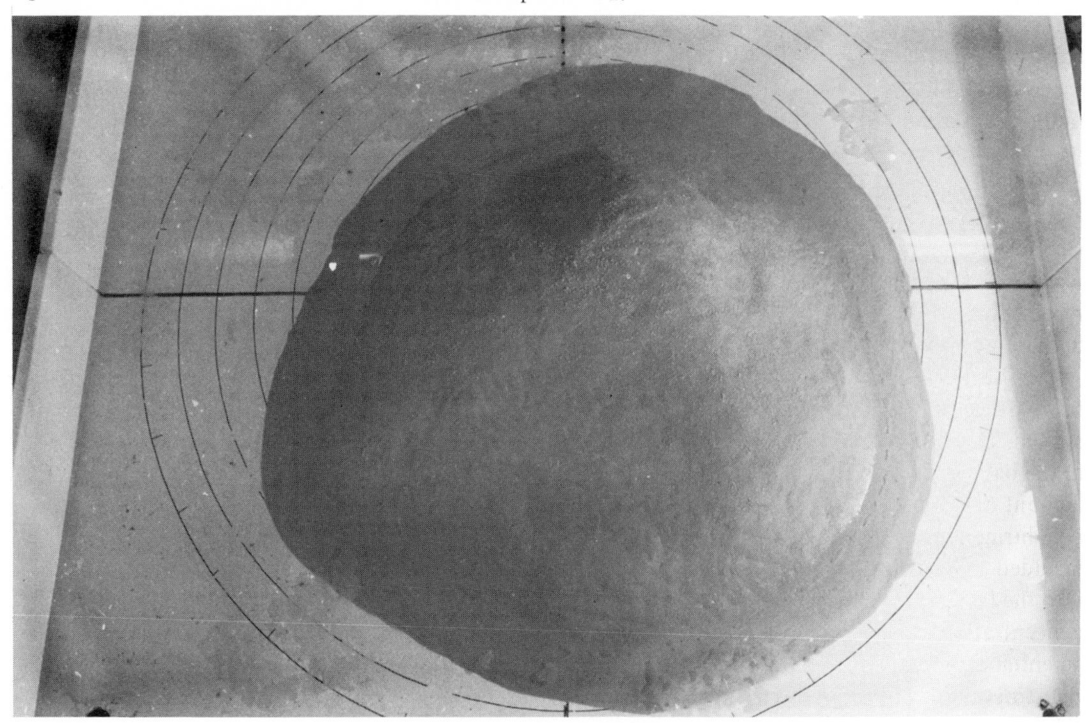

Figure 6.13. Perimeter of island at conclusion of phase 2 of Experiment 1, showing little change caused by deltaic or offshore deposits.

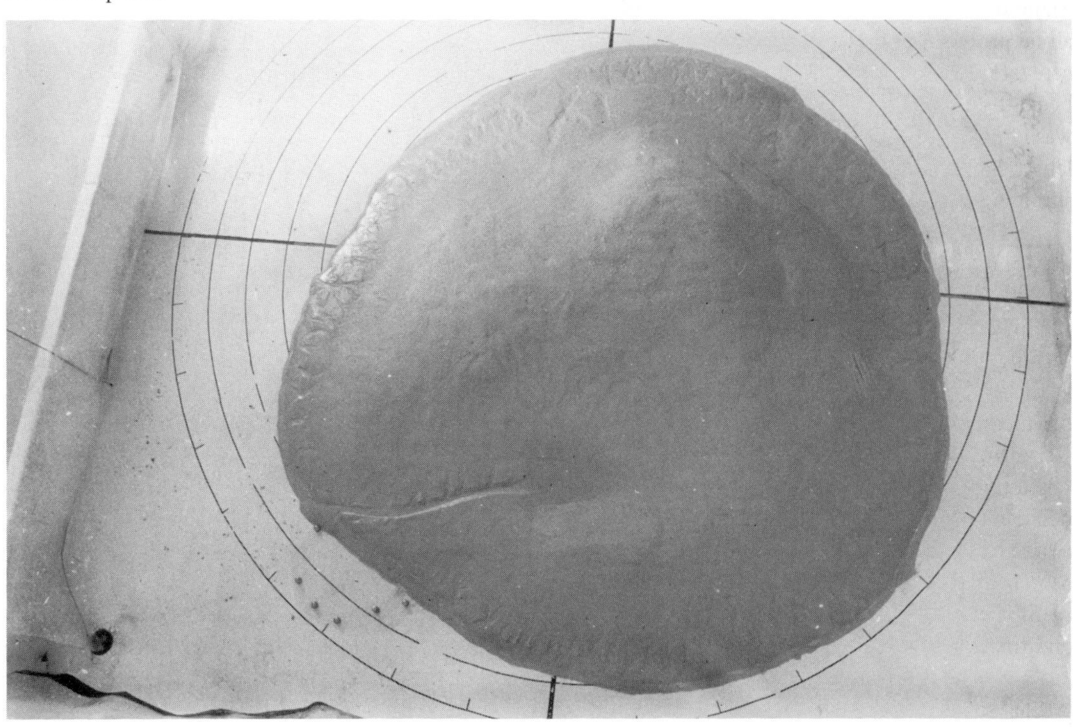

from the extensive northern slopes and joining the main stream at accordant levels, demonstrating the base-level control applicable to Playfair's law.

The main stream functioned as a true stream rather than as a rill, in that it had a clearly defined permanent channel of regular width and depth. It received water by surface flow through tributaries and by basal seepage to produce a considerable discharge with velocity of 2.0 cm s^{-1}. Its catchment was 5240 cm^2 with a total relief of 21.5 cm and is represented by the hypsometric curve (Figure 6.14) with hypsometric integral of 29.8.*

The lake functioned as a sediment trap, with few particles of sand leaving it to enter the stream system. Thus the 33 g of deltaic material deposited in the bay was of fluvial origin and included a small amount of material deposited in the early stream-bed during phase 1. The main-stream catchment included some long rills or incipient rills flowing into the lake, and surface-washed features where concentration of water appeared to pick out differential compression of bed material caused during construction of the island. No measurements were taken of sediment deposited within the lake.

Of the two resistant strata inserted into the island during construction, the steep-slope dyke alone produced discernible changes. Water concentrated behind the dyke and escaped from beneath the blocks in sufficient concentration to wash out sand and cause a failure zone of 17 cm length at the foot of the dyke, accompanied by slumping of about 21 g of sand fronting the dyke.

The island lost 0.5 cm absolute height during the total experiment. It may be surmised that this was due to loss of material and weakening around the lower slopes of the island, accompanied by settling, in such a way that a slight bulge developed at the 6–8 cm levels, indicated by *post-run* azimuth bar measurement.

Note that aspects of the conclusion can be taken up for specific quantification using data collected during and after the run, together with a map of the island.

Systems interpretation: The island erosion experiment is one of the more complex analogue models, and the example used could have been simplified by leaving out the main stream–lake subsystem and the dropping of sea-level. Inclusion of these complexities adds to overall instructional value, but they should not be included unless they are to be incorporated into a total explanation and given due weight. Aspects of the experiment can be separated out and used as the basis for other projects.

In the present experiment, the use of two phases complicates the interpretation of process and form, or cause and effect, and the earlier, constant sea-level system, which reaches general equilibrium conditions during the 38 minutes of running time, is perturbed as soon as the sea-level drop phase commences. The system then remains in a condition of inequilibrium until switch-off time at MINUTE 75 of phase 2.

As the morphological subsystem is common to both phases of the cascading subsystem, it is especially instructive to regard the model as a single system, varying dynamically in time, in the way that field systems usually do.

For the island as a whole there are ideally three sets of morphological data required: the initial landform, the landform after phase 1, and the ultimate landform. The key data are, however, those for the final landform which defines the morphology that has functioned throughout the experiment, unchanged in any major way excepting the addition of offshore slope to land surface.

Figure 6.14. Percentage hypsometric curve for main stream drainage basin.

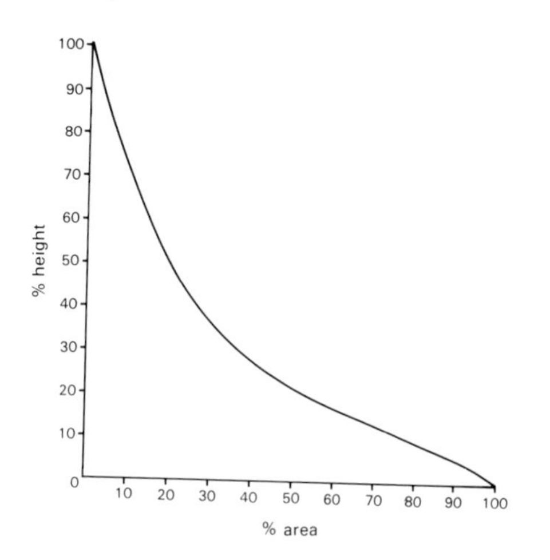

* Using Strahler's concept of % remaining : % original.[2]

Table 6.1. *Morphological subsystem regulators*

Morphological subsystems	Topographic	Drainage net	Regolith	Zone of aeration	Groundwater	Lake
Regulators	Mean island relief, $\bar{H}$ (4.95 cm)	Channel gradient, $\sin\theta$ (0.54)	Permeability, K_r (1.0)	Permeability, K_a (1.0)	Permeability, K_g (1.0)	Cross-section, A_1 (2.875 cm^2)
	Island area, A (19 900 cm^2)	Channel cross-section, A_c (0.75 cm^2)				
	Mean island slope, S (0.2828)	Bed roughness, n (0.01)				

In a model which is concerned with overall processes, as well as with detail, there are a large number of attributes that could be monitored, and there is a need to concentrate upon what is essential and what can be achieved in a limited amount of time.

Attributes that may be considered are:

(*a*) the morphological subsystem (identified by observation and deduction using prior knowledge of processes)
(i) topographic subsystem that delimits total energy input to the island
(ii) total stream/rill/gully subsystem that routes water off the surface to lake or sea
(iii) regolith subsystem that transfers water to the zone of aeration and supplies material to the cascading subsystem
(iv) zone of aeration that routes water to the groundwater body
(v) groundwater subsystem
(vi) lake subsystem that stores water and provides a local base-level.

(*b*) the morphological subsystems incorporate regulators that integrate them to the cascading subsystems, as shown in Table 6.1. The regulators are identified as the attributes of morphology that control quantity and rate of movement of energy and mass through the morphological subsystems, and are usually fairly easily determined from linear measurements or their direct derivatives ($\bar{H}, A, S, \sin\theta, A_c, A_1$) or from tables ($n$) or from other laboratory work (K-values). Of these, determination of K alone involves specialized apparatus.

(*c*) the attributes of morphological subsystems that are linked in such a way that variations in attributes of one subsystem will cause variations in attributes of another subsystem. These are as shown in Table 6.2. The number of interdependent attributes determines the phase-space of the subsystem and is a measure of the subsystem's dynamic relationship with other parts of the total system. These attributes are determined by correlation-analysis based upon measurement during the functioning of a total system. Data of this kind for field studies are scanty. Laboratory models lend themselves to this type of study, but such work is rarely practicable for earth-science demonstration and requires research time, but it is possible to isolate many attributes by logical reasoning and intuition. Values added to the attributes above were already available or were obtainable with little additional treatment of the basic data for the experiment.

(*d*) the cascading subsystems: these relate to the input of water, which moves by a number of processes through the morphological subsystems, loosens and transports debris (sand) for some subsystems and flows or deposits debris into the sea.

The canonical structure for the island system is as shown in Figure 6.15. (Note: loss by evaporation is negligible during 113 minute run.)

Table 6.2. *Morphological subsystem attributes*

Morphological subsystem	Topographic	Drainage net	Regolith	Zone of aeration	Groundwater	Lake
Attributes	Area Relief	ΣL_u	Mantle thickness	Erodibility	Rock porosity	Lake area
	Mean relief	ΣL	Porosity	Rock porosity	Water table	Lake volume
	Relative relief	Drainage density		Rock structure		Cross-section
	Ruggedness	Stream number				
	Mean slope	Number of all streams				
		Channel gradient				
		Channel width				
		Channel depth				
		Hydraulic radius				
		Bed roughness				
		Length bed material				
		% silt & clay				
Phase space	6	12	2	3	2	3

Data used in the above analysis is taken from the end of the total experiment, after dropping sea-level has increased the area of the island and allowed the drainage systems to extend. This is to use values taken from a point in time in a developing model, since none of the subsystems is in a state of equilibrium at time of switch-off (the experiment being stopped arbitrarily for mechanical reasons).

While the total system is perturbed it seems to be unlikely that equilibrium conditions would develop, and feedback loops would have no opportunity to demonstrate their function as they did during the first phase of the experiment.

The major feedback loops in operation are functions of sea-level change and involve attributes in at least two of the morphological subsystems (groundwater and stream network) and arguably in a third (the topographic).

(*a*) Consideration of the sea-level, groundwater subsystem relationship, using the groundwater subsystem H and Q_{total}, where H is the perpendicular distance between crest of groundwater table and sea-level, and Q_{total} is the total effluent discharge per unit time from the island.

Preamble: The drop in sea-level around the island is equivalent to the drop in sea-level around a cone with a slope of about 30°. In such a situation, in applying $Q = KA(H/L)$ to assess discharge through a square centimetre in a unit time, increments of H will be approximately compensated for by equivalent increments of L; that is to say that increase in hydraulic head caused by fall-away of sea-level from the GWT will be approximately compensated by an increase in the horizontal distance through which water has to travel through the ground to seepage points (Figure 6.16). There is, however, the complication of groundwater having to spread laterally across the newly exposed area at the base of the cone, which in turn is in part compensated for by additional rainfall input on the new surface.

The area of incremental coastal strip is, as part of a regular cone, equal to its slant height × its mean length. Slant height and mean length can be determined by direct measurements, but reference to Figure 6.16 shows that slant height (hypothenuse) is related to change in sea-level by the equation

$$\text{slant height} = H \cos \theta$$

$$(= H \cos 30^\circ = 3.5 \times 2$$

in the present example).

The optimum seepage area will therefore be the original sheetflow area, plus the increment area.

The important points are that in referring to changes in H it is not possible to deduce changes in GW-Q with respect to a cone, but it is possible in such a case to relate changes in *total* discharge from the island to changes in H.

As sea-level drops more quickly than water table falls, H will increase by an optimum of 3.5 cm, and total discharge from the island will increase proportionately.

This may be represented by the feedback loop shown in Figure 6.17. This indicates that

Figure 6.15. Canonical structure for island system in Experiment 1.

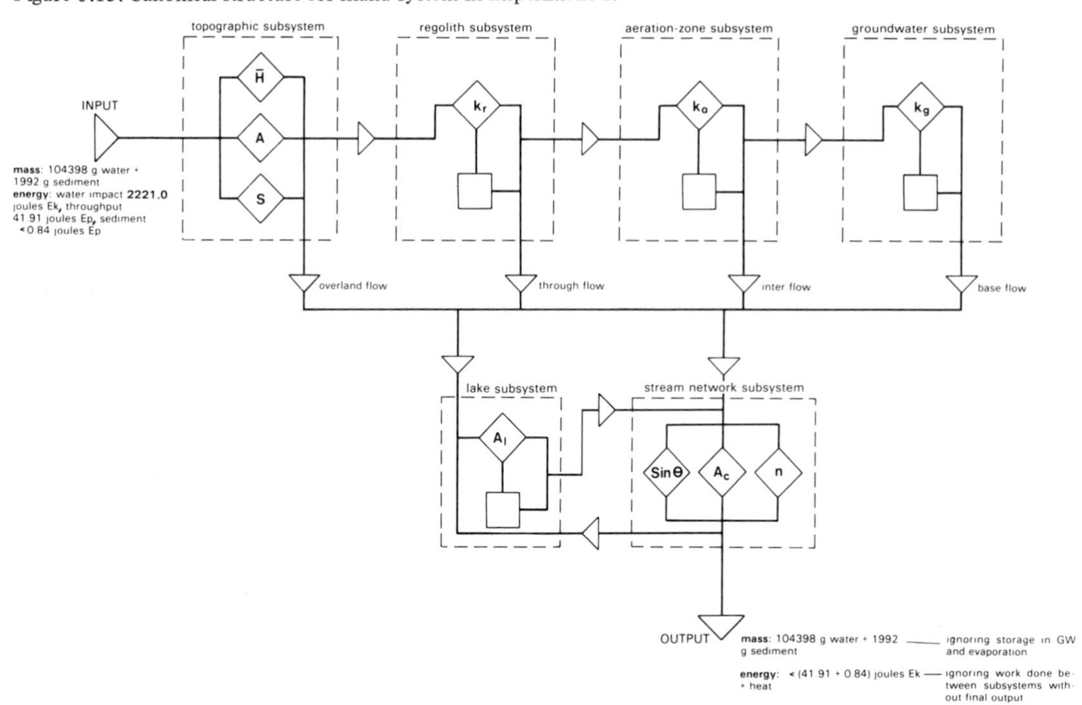

Figure 6.16. Relationship between hydraulic head and lateral flow distance during sea-level drop.

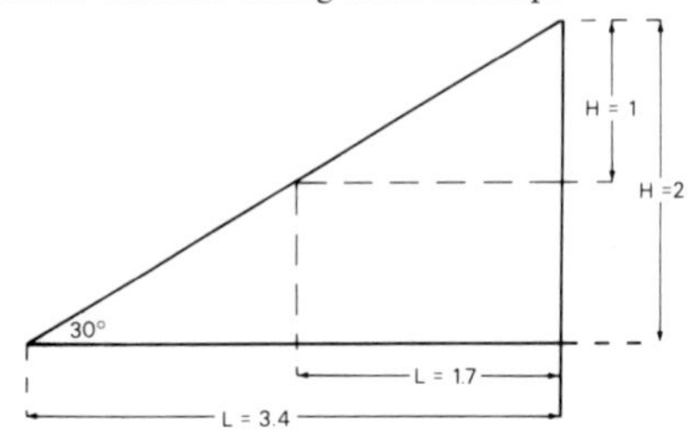

runoff will counteract increase in H, so that a negative feedback loop will develop which tends towards establishing equilibrium between sea-level and H.

(*b*) Consideration of the stream network subsystem in relation to the groundwater subsystem, using the attributes, H, Q_{total}, stream discharge (Q_s), stream width (w) and depth (d), together with surface runoff area from the topographic subsystem. The linkages are as shown in Figure 6.18. The increase in height of groundwater table relative to sea-level will increase surface runoff (seepage) area, which will increase total discharge from the groundwater body and enable streams to extend downward to keep up with the receding shoreline, while losing a smaller part of their length at the headward end due to some increase in the zone of no erosion, X_c. The tendency is for rills to enlarge by extension and piracy, so that their discharge, on average, increases and stream-channel width and depth increase. As a result of increase in stream depth in particular, but affected by a general increase in drainage potential, the groundwater body will be able to drain more effectively and will lose absolute height in the island mass. This

Figure 6.17. Feedback loop (Experiment 1), relating groundwater table and effluent discharge.

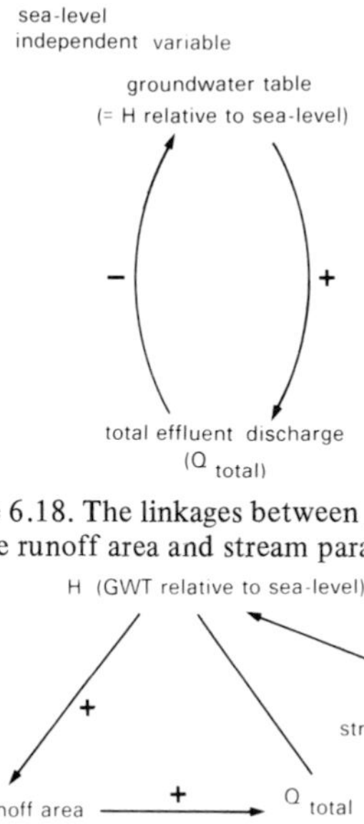

sea-level
independent variable

groundwater table
(= H relative to sea-level)

– +

total effluent discharge
(Q total)

Figure 6.18. The linkages between groundwater level, surface runoff area and stream parameters in Experiment 1.

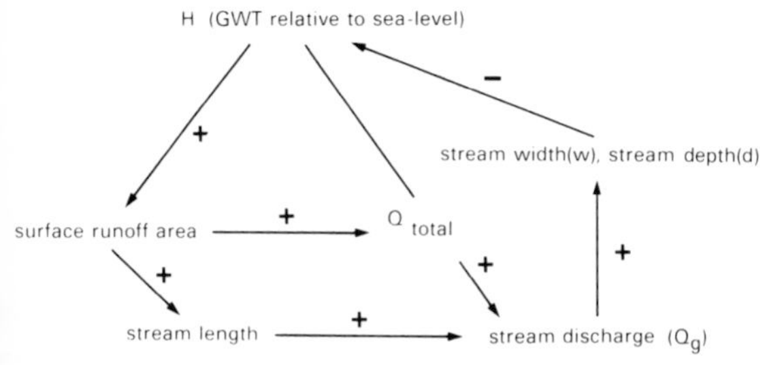

negative-feedback loop would, if sea-level drop were halted, lead to an equilibrium condition in which the various factors were balanced - that is, the factors in $Q = KA(H/L)$, Q_{total}, total seepage area and zone of no erosion, and the stream geometry attributes.

Note: In consideration of the practical difficulties of obtaining velocity values for kinetic energy input in this and other systems, an assessment is made using potential energy. This should be borne in mind when kinetic energy is referred to in analysis (see also p. 38).

Phase 1

Area 12 240 cm^2 (unchanged during phase 1)
Total relief 18.5 cm
Mean relief 3.61 cm ($\bar{H}$)
Duration 38 minutes

Phase 2

Area at commencement 12 240 cm^2
Area at end 19 900 cm^2
Total relief at commencement 18.5 cm
Total relief at termination 21.5 cm
Mean relief at commencement 3.61 cm ($\bar{H}_1$)
Mean relief at termination 4.95 cm ($\bar{H}_2$)
Duration 75 minutes

General Data

Distance from nozzles to hilltop *c.* 200 cm
Rainfall intensity 3.75 cm/hr
Sea-level drop 3.5 cm
Loss in height of island (absolute 0.5 cm)

Phase 1

Input: Mass
= area × rainfall intensity × duration × density of water
= 12 240 × (3.75/60) × 38 × 1 = 29 070 g
Impact energy (E_k)
= (mass × distance)/10 197.17 J
= [29 070 g × (200 cm + total relief − $\bar{H}_1$)]/10 197.16
= 29 070 × (200 + 18.5 − 3.61)/10 197.16
= 612.6 J
Runoff energy (E_p)
= mass × $\bar{H}_1$/10 197.16 J
= (29 070 × 3.61)/10 197.16 = 10.29 J.
(Sediment mass and energy are insignificant in phase 1.)

Phase 2

Input: Mass
= (area at commencement
+ half area increment) x rainfall intensity
x duration x density of water
$= [12\,240 + \frac{1}{2}(19\,900 - 12\,240)]$
$\times (3.75/60) \times 75 \times 1$
= 75 328 g
Impact energy (E_k)
= (mass x height)/10 197.16 J
where *height* is distance of drop fall to hilltop + total relief − phase 1 mean relief − half increment of mean relief during phase 2, i.e.
$200.5 + 21.5 - 3.61 - \frac{1}{2}(4.95 - 3.61)$*
Hence E_k
$= [75\,328 \times (200.5 + 21.5 - 3.61 - 0.67)]/10\,197.16$
= 1608 J
Runoff energy (E_p)
= (mass x height)/10 197.16
where height is
$\bar{H}_1 + \frac{1}{2}(\bar{H}_2 - \bar{H}_1) = 3.61 + 0.67$ cm
$E_p = (75\,328 \times 4.28)/10\,197.16 = 31.62$ J.
Sediment *mass* is 1992 g by collection and weighing.
Sediment energy (E_p)
= (mass x height)/10 197.16
$= [1992 \times (3.61 + 0.67)]/10\,197.16$
= 0.84 J.

Total mass input to island system is water 104 398 g, sediment 1992 g, water impact energy 2221.0 J, runoff energy 41.91 J, sediment <0.84 J.

Note that certain assumptions are made throughout these calculations. In particular it is assumed that sand is derived from all over the phase 2 island surface if mean relief of 3.61 + 0.67 is used, whereas it is known to derive mostly from lower slopes.

* Note that this simple treatment of increasing mean relief is possible if the progressively exposed offshore slope is regarded as a regular cone for which mean relief increases linearly with increase of total relief.

Note also that the separate lake–river system may be treated in the same way. It has parameters: phase 1 area 4129 cm^2, $\bar{H}$ 3.6 cm, H 18.5 cm; phase 2 area at conclusion 5240 cm^2, $\bar{H}$ 4.5 cm, H 21.5 cm. This data cannot be worked into Figure 6.15 which recognises only the lake as a subsystem, rather than a subcatchment.

As stated at the beginning of this chapter, attention should be paid in analogue models to the use of terms which have a size or process connotation. These are: sea, shore, offshore slope, mountain or hill, rill, gully, stream, lake, sill, dyke, drainage basin or catchment, delta, channel, tributary network, groundwater body and groundwater table.

If the model were run as a micro-system without scalar properties, it would still demonstrate some very important processes and concepts, but some of the terminology would be inappropriate - and, indeed, for micro-systems, no scientific terms exist with which to identify features. From the list above perhaps only the terms rill, delta, channel, tributary network, groundwater body and groundwater table could be used as easily for micro-forms as for field examples.

If, as was done, the model is taken to represent a prototype on a scale of 1:10 000, then the terminology used is acceptable, but only as applied to the prototype that the model represents. That is to say the terminology is largely concerned in this case with an idealized reality. In fact there remains a lack of appropriate terminology for the model itself. This can only be overcome by removing, in many instances, a size connotation from the terms. In some cases a process connotation also exists.

The processes in the model are probably similar to processes as they would be in an idealized prototype with the same controlling attributes. The nature of these processes is erosion by dislodgement and transport in rills, erosion by sapping and slumping in groundwater locations, and deposition in channels, lake and sea. Since there is no common usage for terminology in these processes they may be referred to in this model or its prototype without concern for scale or misrepresentation. This matter is discussed further in Chapter 12.

7 The flume in instructional contexts

The processes and general characteristics of stream development are best studied in a flume. This apparatus enables better control to be exercised over the position, magnitude and development of the stream than is possible with the rainfall simulator. A point source is employed instead of a general rainfall, and water in the stream channel remains more or less constant in quantity throughout its length.

The flume stream channel - thought of as being within a mobile sand bed - receives most, if not all, of its input from upstream at the head, from a headpool feed system, so that the segment is representative of a much longer stream, the origin of which is morphologically beyond the flume itself. This is in contrast to the streams that develop in a rainfall simulator, which are hydrologically, hydraulically and geomorphically complete systems. Flume stream segments are therefore more uniform in channel characteristics, and may be very much larger than simulator streams.

The design of a flume is flexible, and the general plumbing requirements are discussed in Chapter 4. There are, however, three essential components of flumes - they have a box-like container for the stream model, a controllable input at the head, and a controllable outflow at the foot. A variety of sizes and construction techniques exists: a simple 81.3 cm × 25.4 cm × 8.9 cm deep model; a 2.95 m × 0.81 m × 0.20 m deep model; and the more complex flumes of commercial design sometimes used in university and research laboratories.[1,2,3,4] It has been found that the type of glass-sided flume used in sedimentation and flow studies, commonly 1 ft square in section with a very high flow potential, is unnecessarily powerful, not ideal for a wide range of experiments, and difficult to adapt.

Preference is given to a flume of 3.5-4.9 m in length for earth-science experiments, and 10 m flumes are not uncommon. Wooden construction has much to commend it - being easy to construct, relatively cheap, flexible in design potential, and readily modified. Flumes of 1-1.25 m in length are useful for a range of elementary demonstrations but may prove to be frustrating for more sophisticated work.

A 6 m flume will be described, but the principles of design and construction apply equally to small bench models. When in use, a flume may contain a considerable mass of water and sand; for instance, if a flume 6 m long by 61 cm wide contains water to a depth of 9 cm, it is carrying well over 550 kg. Consequently, the structure of flume and stand must be robust, and this is even more important if it is necessary to be able to tilt the flume, because of torque and distortional strains that will be set up in the apparatus. The hinged-type model is particularly weak and distortion may be detected in accurate work.

A flume is best set on its own stand so that its upper edge is about waist high, to facilitate setting up experiments, monitoring and observing. The stand may be of wood (5 cm × 10 cm studs) using a triangular frame to strengthen the legs. To facilitate moving the apparatus, it is advisable to make the base in sections.

If the base is made of metal, it is most likely to be of 2.5 cm or 3.8 cm angle-iron in a box-frame construction, with heavy bolts in the floor members to permit levelling and tilting (Figure 7.1). Tilting a large flume is awkward, requiring a datum line to be stretched from head to foot to indicate the position required.

Flumes that are hinged at one end, or at a point along the flume floor, are so designed to produce a sloping test bed. It has been found that such a technique is rarely necessary - sloping beds can be achieved in two simpler ways, as is described below.

The flume will have to be rigid enough to retain its shape when containing sand and water, and it will be found that the depth of the flume is critical in this respect. If a deep flume is used (for example, a 9 m × 0.61 m × 0.61 m flume built by the author at the University of Cape Coast) water pressure at the bed will be so great that sealing will be a real problem. With 46 cm depth of water, pressure at the bottom will be about 46 g cm^{-2}, which puts a considerable strain on the bottom, sides and joints: the sides are likely to warp and require reinforcement (Figure 7.2), and the joints will tend to spring.

A 25 cm deep flume is generally more useful, and 7.5–15 cm depths may be adequate. The sides may be of 12.5 mm marine plywood, cut from 2.44 m × 1.22 m sheets and carefully trued to bond with the floor. It is advisable to screw the sides onto the top of the floor, since the floor will be supported by the base, and the pressure on the sides should set up a shear stress on the screws rather than drag them from their holes.

Sides of 12.5 mm plywood will tend to warp and brackets will have to be incorporated into the design at 1.2–1.5 m intervals. These may be of angle-iron (Figure 7.1) or wood (Figure 7.2). These brackets should be at least 7.5 cm short of the top so that they do not get in the way of apparatus sliding along the top of the flume.

Joints must be bonded with a good waterproof wood glue, and the whole should be sanded inside and out to remove rough patches and splinters.

If funds are available, a much stronger flume is produced from 5 cm thick wood. This will not warp in use but requires more effort in construction. The flume in Figure 7.1 was constructed by the author with ordinary hand-tools, and a technician later added to it. The angle-iron base was made by the technical service department.

In demonstration experiments, very high discharges are infrequently used and a delivery maximum of 150 $cm^3 s^{-1}$ will often suffice. Some experiments described use more than this but it is

Figure 7.1. Base of flume, showing box-frame construction and bolts in base to permit levelling and tilting.

not normally essential that they should. Plumbing should be able to carry water in excess of the optimum discharge, particularly at the outlet, since this is the least adaptable part of the apparatus.

A re-circulating system should be used, which consists essentially of a constant-head feed reservoir suspended above the head of the flume (Figure 7.3), an outlet pipe which permits free-fall of water through an adjustable pipe into a catching tank, and a return pipe to a main reservoir tank, which in turn supplies a pump.

The pump need not be as powerful as a rainfall simulator pump - a small $\frac{1}{8}$ h.p. centrifugal pump has been found to be adequate for a wide range of experiments. Alternatively, it is possible to use one pump for both simulator and flume, if laboratory space and pipe connections can conveniently be adapted. Water is pumped into the feed reservoir (Figure 7.4), and a constant head is maintained by one or more large overflow pipes (>3.8 cm $= 1\frac{1}{2}''$ internal diameter) leading back into the main reservoir. The higher the feed reservoir water surface is above the feed tap, the greater the pressure, and better the control, so that a deep feed reservoir (*c.* 25 cm deep) is advisable, and to facilitate catching feed-water to measure input, the reservoir should be poised at 38-50 cm above the flume. If an *organ-pipe* feed-system is used, then the depth of the feed reservoir is less important.

One of the main problems in flume design is the control of input to the headpool, that is to say, the control of stream discharge or throughput. Taps or cocks normally have poor fine adjustment, and some designs are almost useless for sustaining an accurate flow. This is particularly so when very small inputs of, say, $10\,cm^3\,s^{-1}$ are used, but even large values will be found to creep out of a required range during a run. The worst type is the cone or disc plunger that seats into a circular orifice in the tap (Figure 7.5). At low deliveries this type is almost impossible to control because of a rapid transition from zero to moderate discharge. Further, taps and cocks suffer generically from the build-up

Figure 7.2. An all-wood flume with side reinforcement.

Figure 7.3. Head of flume, showing constant-head feed reservoir.

of air bubbles, due in part to the warming of water as it passes through the brass piece, and in part to the lodgement of air at the end of the previous experiment as water recedes up the tap interior. Air bubbles are normal at low discharges but occur frequently at high discharge also. The tap of Figure 7.5 is more subject to bubble problems than most.

A somewhat better tap is a simple shaft-in-a-tube model that turns on when the hole in the shaft is aligned with that in the tube (Figure 7.6). This is the burette tap principle and is commonly found in dairy fittings. It is relatively free from bubble effects, permits a fine flow of water, but is difficult to adjust.

Gate-cocks are capable of fine adjustment and good flow control, since the gate is slid over the hole by a screw action and can open up a very fine aperture. They are, however, subject to bubble formation and need to be force-flushed before a run. If a very small input is required, an actual

Figure 7.4. Feed reservoir with inlet pipe to left and weir surrounding the return-pipe outlet to control accurately the standing water level.

Figure 7.5. Tap with disc plunger.

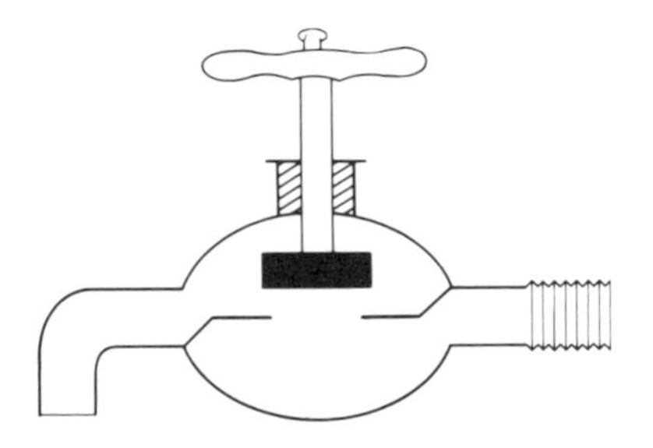

Figure 7.6. Shaft-in-a-tube tap.

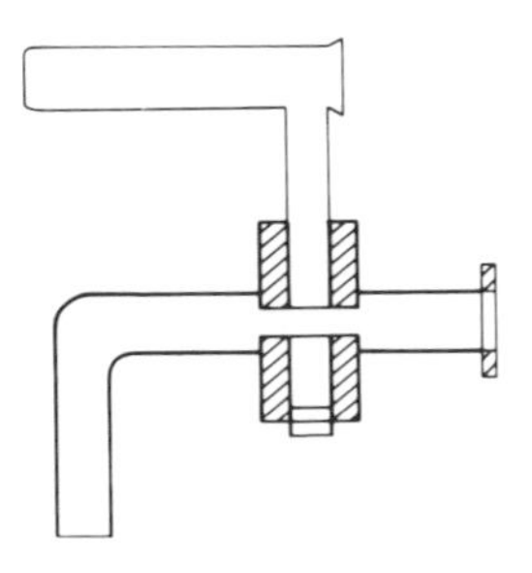

burette tap can be fitted to the output from an ordinary gate-cock, which is used to cut down the supply to approximately the correct value.

It is sometimes necessary to connect the tap from the feed reservoir to the actual point of input to the headpool by rubber tubing, to avoid splash and agitation, and if this is done, special care must be taken to avoid airlocks. A length of rubber tubing has the advantage of permitting a pinch clip to be used to cut off flow, so that the tap setting may be left alone for a future experiment.

A very satisfactory feed-system has been developed in the Memorial University laboratory, making use of pipes which will slide up or down through a stuffing box to put them into or take them out of action. These can be of varying diameters, each with an oblique cut at the top, providing a parabolic section for water to enter the tube and drop straight down into the flume. This *organ-pipe* feed-system is free from all the defects of taps and cocks, such as airlocks and metal expansion, and permits highly accurate inputs to be achieved and maintained (Figure 7.7).

Figure 7.7. Organ-pipe feed-system.

A draining tap should be fitted to the flume trough at the head, near to the floor of the flume, to allow draining of the headpool without disturbing models set up in the body of the flume. This may be of any adjustment type, but must have a flanged and threaded shank so that it can be fitted into the flume wall. If the wall is too thick for the threaded shank, a plywood inset can be used which will be waterproofed on the inside by rubber washers.

At the foot of the flume a similar tap may be set into the flume wall, but the main outlet should be a free-fall pipe. Removing water from a flume is problematical, since pumps are difficult to control precisely, and are of little use. Siphons are usually too slow and are difficult to control; taps are prone to develop air bubbles and are difficult to adjust finely. Weirs are highly effective and offer an accuracy of control similar to that of the sliding pipe, but they are not so readily adjusted.

Figure 7.8. Main outlet from flume.

As in feeding a flume, therefore, the free-fall pipe is the best device (Figure 7.8). This should be of 2.5 cm–3.8 cm metal or plastic pipe, smooth and polished and set into a stuffing box through which it can be slid up or down to control the level of sea-water. The stuffing box grip can be adjusted by tightening the lower cap (Figure 4.8), so that the pipe is held firmly in any position but can be slid up or down by hand. The upper flange should be as thin as possible to permit low depth of sea-water, and the outlet pipe must be kept free from sand grains and periodically greased.

Sea-level is controlled by the position of the outlet tube and water must be allowed to fall freely down it, leaving a suction cone with an air-core in the tube, otherwise the sea-level can rise slightly and control is lost. The water falls into a catching tank beneath the flume (Figure 7.9), which is connected to the main reservoir tank by a 5–7.6 cm internal diameter pipe, which should be as free from bends as possible.

The water level in the catching tank will be at about the same height as that in the reservoir tank, and it is important that it should be able to drain freely. For this reason it is best to make the catching tank as tall as possible so that, if water falls into it very rapidly, a hydraulic head can build up and increase the flow rate to the main reservoir. In any case, the height of the catching tank should be no less than that of the main reservoir, and the connecting pipe should be of as large a diameter as possible. The rule is that water moving under gravity requires wider pipes than that under pump pressure, or water moving under low hydraulic head requires wider pipes than that under higher hydraulic head, in order to maintain a constant, balanced flow throughout the system. Piping in small flumes or stream-tables is usually inadequate; this is their main fault and it is particularly significant in their outlet system.

The reservoir tank could be used as the catching tank also, but it is too large to fit easily under the flume, or to clean and service, if tucked away. Also, a choice has to be made between using a flume-length pipe to return water from the catching tank to the main reservoir, and a long pipe from the constant-head feed reservoir overflow to a tank at the foot of the flume. In practice, maintaining a constant head in the feed reservoir is more critical than returning water from the outflow at the foot of the flume to the main reservoir. The former is best done by a wide-diameter, short pipe, or even a short chute, which means locating the main reservoir separately, near the head of the apparatus (Figure 7.10).

The flume can be lined with 1000 gauge polythene sheet, care being taken to avoid crinkles. It should be fitted and fixed before holes are cut for piping. This type of lining can last for years if care is taken not to damage it with tools during model formation and servicing. A better lining is produced with fibre-glass. In either case, the flume should be of white finish with waterproof calibrations where necessary. Attempts at caulking and sealing wooden flumes with varnishes or pitch have always been unsuccessful. The water pressure from within tends to spring open joints, rather than as in boats where water pressure tends to seal the joints.

There are a number of controls and measurements so often used in flume experiments that it is advisable to incorporate basic calibrations. The feed-system depends upon pump pressure, and although the constant-head tank will allow excess water to escape, it will be found that when low deliveries to the flume are used the overflow pipe may be over-taxed, water level may rise in the feed tank, and control is lost. This problem may be avoided by putting in a pressure-reduction cock between pump and feed tank to reduce flow, and this may be monitored by inserting a 1750 g cm^{-2} (172 kPa or 25 psi) pressure-gauge between valve and feed tank.

Delivery taps into the flume head may also be approximately calibrated by a circular dial, set beneath the handle and marked off in commonly-used Q-values.

When a model is set up in the flume, it frequently separates the headpool from the sea, and these will maintain different levels. The headpool

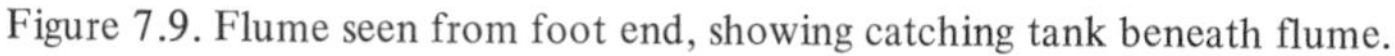

Figure 7.9. Flume seen from foot end, showing catching tank beneath flume.

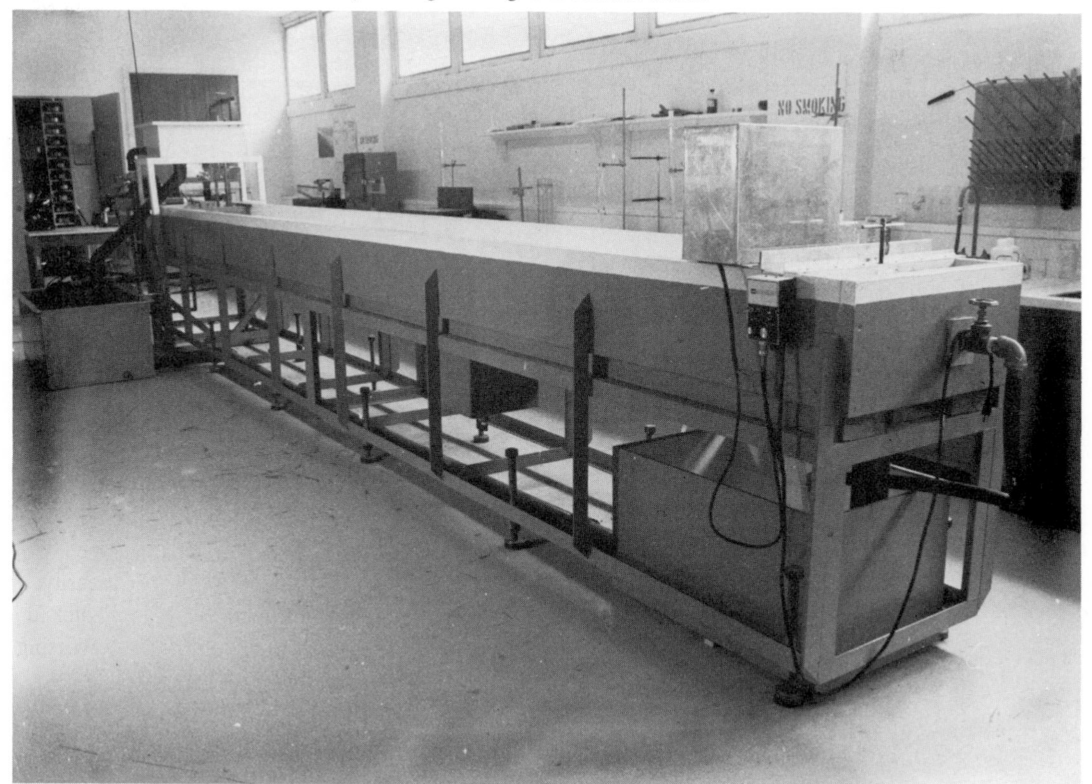

Figure 7.10. Main reservoir tank, at head of flume.

requires markings on the flume wall - a tape-gauge - in 0.5 cm intervals: intermediate values can be estimated. Sea-level may also be monitored by a tape-gauge, but a more accurate device is a float-gauge which may be constructed as in Figure 7.11. This consists of a plastic disc cemented to three ping-pong balls, with a marker rod sliding through two ring hooks. It is easier to read than a tape-gauge, which suffers from meniscus refraction and, quite often, scum films.

If small movements of sea-level are to be monitored, the float-gauge can be attached to a lever-type multiplying arm which is read off against a circular scale.

The outlet pipe can carry its own gauge for convenience in pre-setting sea-level (Figure 7.12) or, alternatively, the sea can be run in to the level required and the outlet pipe is then pushed down until the sea just begins to flow out from it.

The top edge of the flume should be marked off from the head end, starting at zero, in metres, decimetres and centimetres. This scale will be on the trued edge of the flume, and it will be subject to wear by the sliding carriage and other equip-

Figure 7.11. Float-gauge for monitoring sea-level in flume.

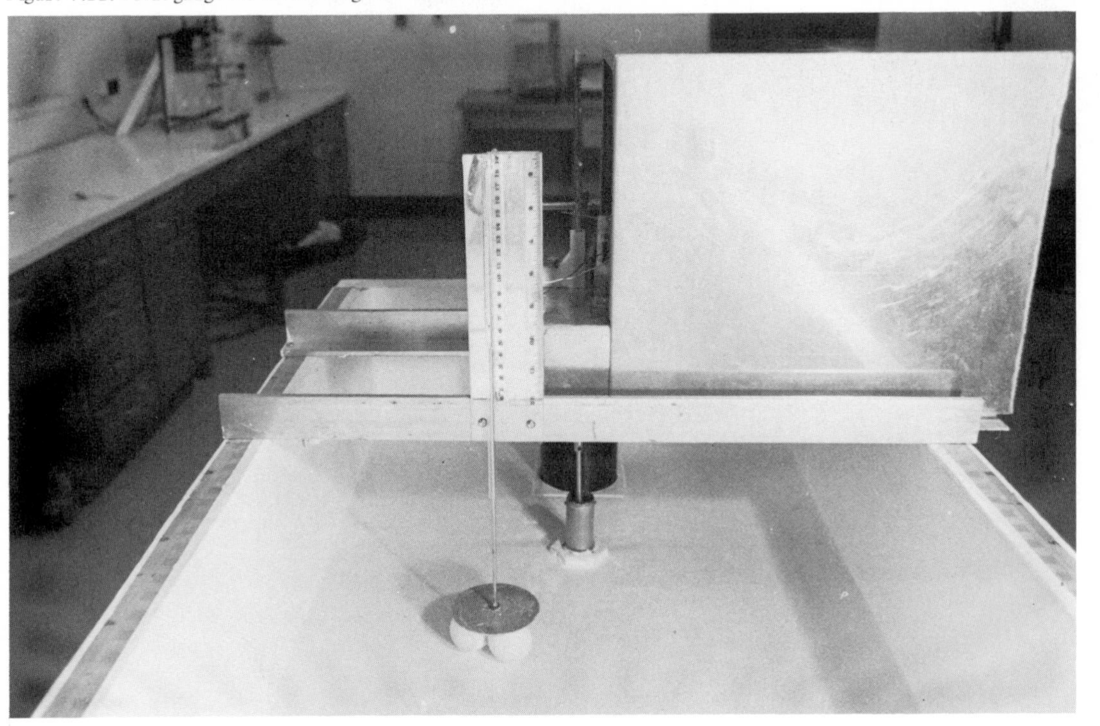

ment, so it should be protected by a good silicone varnish.

The main reservoir tank should be large enough to contain all the water from the system, and the catching tank at the outlet should be tall enough to allow the main tank to fill without itself overflowing. Accordingly it is advisable, when filling the system, to introduce sea-water into the flume from the main tank via the pump, disconnected from the constant-head tank, to ensure that too much water is not added to the system. In other words, using a hosepipe from a sink-supply to fill the flume should be avoided until the holding capacity of the system is well known and can be properly judged. It is also best to use water at the ambient laboratory temperature to avoid bubble formation in the feed system.

Before the flume is used for an experiment, the feed-system must be set. The feed tap should be turned on to deliver approximately the required input, and the pump delivery should be adjusted until the constant-head tank overflow pipe is functioning correctly. The feed tap is then carefully adjusted by collecting measured volumes of water in stop-watch measured time intervals. Once the correct input is obtained the flow is stopped, by switching off the pump or plugging the delivery pipe in the tank with a rubber bung. The tap setting should be left alone. If the pump is switched off and the feed pipe empties completely, it may be necessary to remove air bubbles by forcing water backwards through the tap before an actual run is commenced. The flow should be checked before, during and after a run.

The sand to be used in the model is put into the flume at some distance down from the head, say 1-2 m, leaving a large headpool to smooth out ripples and bias from the input to the model. This sand will have to be roughly modelled into the required bed shape, and then thoroughly wetted. If the sand is freshly added, it is best to run in the sea and headpool water in order to saturate the sand bed thoroughly, then model it somewhat more accurately and shock it by pummelling with

Figure 7.12. Outlet pipe with gauge.

the edge of the hands or with a fairly heavy stick. This drives out air cavities and excess water and allows the sand grains to cohere firmly. If this is not done, the sand may look firm but will behave as a sludge, and results will be poor. Once compacted, the sand may be scraped, cut and formed and will retain the most exacting shapes. It tends to improve in character in this way over a period of days.

The sand can be given the final required shape when the headpool and sea have been removed again, by running the carriage over the top of it with the appropriate blade attached (Figure 7.13). A wide variety of initial surfaces can be produced in this way, and for the sake of replicability the models should be formed free-hand as rarely as possible.

When the top, frontal and rear surfaces have been formed and trimmed, and the excess sand removed from the sea and headpool floor, the sea and headpool water are reintroduced. This has to be done with great care to avoid spoiling the end faces of the model. Also, the headpool and sea-levels will have to be maintained in rough equality during filling, to avoid the possibility of base-flow through the model causing outwash by seepage. Unless the end faces have been pre-shaped to an underwater angle of rest of about 35°, slumping will occur as sea-water and headpool water rise, and this must be taken into account in the calculation of model dimensions. The technique is therefore: (i) set water input; (ii) emplace sand and roughly mould; (iii) run in sea-water and headpool water; (iv) mould model bed more carefully and

Figure 7.13. Sliding carriage with former blade attached.

Figure 7.14. Irregular water table caused by insufficient standing time.

shock to firmness; (v) remove all water from flume; (vi) shape up model to required accuracy, trim, clean up flume; (vii) carefully reintroduce sea and headpool water.

The model surface should be at least a centimetre above sea-level at the commencement of the experiment in order to avoid the possibility of seawater spoiling the model surface. This is an arbitrary precaution, since some experiments require flooding of the model surface, and some require relatively high relief. The headpool functions as a large, smoothed-out water-supply to the model, and it characteristically enters a pre-formed stream channel. Its level must therefore be carefully raised until it is able to flow into the head of the channel, but not flood over the surface of the model. This combination of water input, channel section and model surface must be thought about beforehand. The initial channels must be of the right size to carry a certain input without flooding. Mock runs can be used to test out a required combination until experience guides experiment design.

If the model includes a river channel, it will be found that unless the sea and headpool have been left in position for some hours - ideally overnight - the water table will be concave, with water seeping in from both ends of the test bed (Figure 7.14). Consequently, when the river begins to run it will lose water by influent seepage for some time before the water table and stream surface are coincident and the stream flows into the sea.

Loss of water from the flume by evaporation has been found to be negligible during most experiments, but if a run lasts for a considerable time evaporation loss must be taken into account. It may be estimated by taking the water input from the constant-head reservoir and the water output from the outlet pipe. More accurately, it should be measured over the sand and open water surfaces separately.

8 Experiments with the flume

There is an infinite number of experiments for which the flume may be used, including river, general landform, coastal, sedimentation and groundwater experiments. In this chapter three types of experiment are considered, namely the channel-process, the general landform and the groundwater types.

The literature on practical river studies is extensive, but an excellent working guide is Blench's *Mobile-bed fluviology*,[1] and works that unite a practical and theoretical approach to river processes are Leopold, Wolman and Miller[2] and Gregory and Walling.[3] Schumm[4] offers a broad perspective in interpretation of process, and Melhorn and Flemal[5] include chapters on themes of particular value in systems analysis while paying tribute to G. K. Gilbert, whose early work with flumes is still inspiring. Morisawa[6] approaches the dynamics of streams at the undergraduate level, and no better treatment yet exists for systems interpretation in earth-science phenomena than Chorley and Kennedy.[7]

8.1 Dimensionless parameters

In demonstration work, providing processes are truthful, the scalar properties of models as compared with field examples will not be of primary importance, since a scale of 1:1 can be assumed. Examples of some of the similitude factors to be considered when field examples are to be compared with models are given for the rainfall simulator experiments in Chapters 5 and 6, for the flume in Experiment 9 of this chapter, and for field models in Chapter 12.

With flume models in particular it is possible to consider the dynamics in some detail with readily obtained data - that is to say, data collected with probes, metre-sticks, stop-watch and dye. These hydraulic parameters are used in maintaining the process truth or effectiveness of models.

Dimensional parameters that control flow mode and the behaviour of water in streams are: the bed roughness (a value denoted by Manning's n and obtained from published tables); the velocity, V of the stream; the slope of the water surface (using the sine of the angle of slope, S); the hydraulic radius of the channel (area/perimeter, R_h, which in a shallow stream is equivalent to depth, D); the kinematic viscosity of water (absolute viscosity/density, υ, which is temperature dependent and may be taken as 0.01 $cm^2 s^{-1}$ in a warm laboratory).

These properties of streams in channels are connected by several dimensionless relationships which give a quantitative guide to flow modes.

In calculating modes from the experiments described in this book, it should be borne in mind that collection of dimensional information is crude; in particular, stream depths are measured with probes for which surface-tension effects are of the same order of magnitude as the depths (often only two or three millimetres), and velocity is measured with dye between fixed points and the velocity of the fastest thread is evidently higher than the mean velocity. With these constraints in mind, the flow characteristics of a typical laboratory stream can be examined.

Using Experiment 1 and extracting information from the *Control* and *Observation* sections, measured velocity, V is 34 $cm\,s^{-1}$; water depth, D is 0.2 cm; water slope, S is approximately the same as channel slope, and using only the last 53 cm from the nick-point is 1.6 in 53 = 0.03; bed roughness, n of medium sand is about 0.011; discharge, Q is 20 $ml\,s^{-1}$; stream width, w is about 3.5 cm; kinematic viscosity, υ is 0.01 $cm^2 s^{-1}$.

Considering stream velocity, there are three ways of assessing its value. The measured value by dye is 34 $cm\,s^{-1}$. The value by the equation

$Q=wDV$, $V=Q/wD=20/(3.5 \times 0.2)=28.6 \text{ cm s}^{-1}$, and by the Manning formula

$$V=\frac{R_h{}^{2/3}S^{1/2}}{n}$$

or

$$V=\frac{D^{2/3}S^{1/2}}{n}$$

$$=\frac{0.002^{2/3}\,0.03^{1/2}}{0.011}$$

$$=\frac{0.016 \times 0.173}{0.011}.$$

Hence $V=0.252 \text{ m s}^{-1}=25.2 \text{ cm s}^{-1}$. (Note that D is in metres.) Of these, the dye-measurement is understandably high. Both the other values, 28.6 and 25.2 cm s^{-1}, depend upon the accuracy of depth measurement, but the equation $Q=wDV$ also uses the dependable parameters of discharge and stream width, while the Manning formula uses a fairly crudely estimated slope factor. It is best, therefore, to accept 28.6 cm s^{-1} as the mean velocity in this instance.

With low discharges, such as 20 ml s^{-1}, laminar flow can be expected in many situations and in Experiment 1 laminar flow was observed to occur upstream where water surface slope was minimal and velocity was very low. Downstream, however, flow was observed to be turbulent. The theoretical check on laminar or turbulent flow introduces viscosity and the dimensionless Reynolds number, R, defined by $R=VD/\upsilon$.

In Experiment 1, $R=(28.6 \times 0.2)/0.01=572$. (Note that here D is in centimetres, since V and υ are in centimetre units.)

The threshold value between laminar and turbulent flow is not precise, but in an open channel is about 500. The flow, therefore, is just in the turbulent mode and may be associated with bed erosion and sediment transport. Upstream, where V decreases, $R<500$ and laminar flow occurs, unable to erode or transport. In the fastest thread of downstream flow R will be a little higher than 572.

In order to take the effect of gravity into account the dimensionless Froude number, F, defined by $F=V/\sqrt{gD}$ is used, which relates velocity, depth and gravity. The Froude number is generally small. When $F<1.0$ the flow is tranquil or subcritical, and the bed may be smooth or have small, more-or-less stationary ripples. When $F=1.0$ the flow is critical and standing waves develop on the stream surface, and when $F>1.0$ standing waves may persist and antidunes may develop which appear to travel upstream.

In Experiment 1, $F=28.6/\sqrt{981 \times 0.2}=2.04$, which is a value for supercritical or streaming flow. High Froude numbers are not often associated with the experiments in this work, but occasionally a channel will narrow for a time, velocity will increase considerably, and shooting flow with a high F-value will occur. Very often, segments of streams will be in tranquil flow with $F<1.0$.

If a stream is to model a field prototype, the Reynolds and Froude numbers should be of the correct order of magnitude. This will usually necessitate distorting the geometric parameters.

As with other laboratory work, a semi-formalized data sheet helps to keep a check on essential data collection and observations during an experiment. This is particularly important during the running of dynamic models of earth-science type, since the change in form and process may be rapid and it is often difficult to select key parameters while observing the whole. A specimen data sheet is shown in Figure 8.1.

8.2 Channel processes

Experiment 1

Purpose: To demonstrate the development of a channel segment and to relate process and form.
Apparatus: Flume 20 ft x 2 ft x 1 ft deep with constant-head water feed facility and adjustable outlet; medium sand formed into a test bed 1.17 m long, 55 cm wide and 11.0 cm deep; sliding carriage with straight former blade plus channel scribe; various scrapers, brush, measuring cylinder, timer, balance and glassware.
Procedure: The water input was set. The sand bed was formed roughly in the body of the flume at sufficient distance from the head of the flume to produce a large headpool. The sand block was wetted by introducing the headpool and sea-water;

Figure 8.1. Specimen laboratory sheet for experiment with flume.

Laboratory Sheet

Flume Experiments Date

Objective

Apparatus

Procedure

Data

Start time ______

Finish time ______

Duration ______

Discharge ______

Sea level at start ______

Sea level at finish ______

Sea level drop rate ______

Sea level change ______

Headpool ______

Bed material ______

Channel and valley width and depth:-

Distance from headpool										
Width of Channel										
Depth of Channel										
Width of Valley										
Depth of Valley										

Delta

Length ______

Width ______

Wet Volume ______

Dry Mass ______

Channel or valley curvature: ______

Other features

Form:

Process:

Conclusions

Photos

Frame No.

1 ______
2 ______
3 ______
4 ______
5 ______
6 ______
7 ______
8 ______
9 ______
10 ______
11 ______
12 ______
13 ______
14 ______
15 ______
16 ______
17 ______
18 ______
19 ______
20 ______
21 ______
22 ______
23 ______
24 ______
25 ______
26 ______
27 ______
28 ______
29 ______
30 ______
31 ______
32 ______
33 ______
34 ______
35 ______
36 ______

the sand was shock-hardened and somewhat better shaped with the carriage former. The sea and headpool were emptied; the test bed was finished, trimmed and brushed clean; the flume floor was cleaned up and the delta apron emplaced (Figure 8.2). The headpool and sea-water were slowly added. The front end of the model was allowed to slump out freely to give the offshore slope (Figure 8.3). The sea-level and headpool water were carefully adjusted to give about 1 cm drop between the head of stream and the sea-level (by first building up the headpool until it just lapped the stream channel, and then bringing the sea-level to within 1 cm of the channel floor at the mouth by manipulating the outflow pipe). When all was ready, the stream was allowed to run and the timer was started as channel flow reached the sea.
Controls: (Scale 1:1) Input to stream 20 ml s^{-1}; sea-level 9.3 cm, constant throughout run; headpool 10.3 cm at start but 10.9 cm during run; duration of run 55 minutes.

Observations and data: Once the pre-formed V-groove became a water conduit, it assumed a flat-bottomed, steep-walled channel towards the sea (Figure 8.4, MINUTE 8.5). The stream velocity was of the order of 34 cm s^{-1}, giving a laminar flow mode upstream and a turbulent mode in the lower segment.

With constant sea-level, the stream graded back to a nick-point which retreated rapidly up-channel for the first 14 minutes, and then at a slower rate until the end of the run at MINUTE 55 (Figure 8.5). Lateral erosion and a flat bed occurred along the whole length of the stream (Figure 8.6), but the optimum lateral erosion occurred downstream once the nick-point had passed, and thus had occurred at the end of the run up to 53 cm from the mouth.

The progressive widening and deepening of the channel was monitored at a position 7 cm from the mouth to give the values shown in Figure 8.7. Widening occurred by a process of bank under-

Figure 8.2. Test bed for Experiment 1, shown from foot end of flume.

Figure 8.3. Test bed for Experiment 1, showing slumping at front end after introduction of sea-water.

cutting and slumping, which occasionally effectively narrowed valley width by partial slumping of an overhang.

Some differentiation of the stream bed occurred, as may be seen in Figure 8.6, to produce incipient pools and riffles at approximately 30 cm intervals. Material from the bed and side was carried down to the sea, mainly by saltation, to be deposited in the delta (Fig. 8.8). The delta yielded 309.8 g of oven-dried sand.

Conclusion: The stream channel developed to a trapezoidal cross-section under the basic controls of sand size and density, permeability, water input and slope of water table. The water table must have had a slope of 1.6 cm in 1.17 m = 1:73 or 1.4%, and this enabled the stream to erode vertically in the downstream 53 cm segment until cessation of the experiment at MINUTE 55. This degrading appears to have occurred in two phases, the first indicated by a nick-point recession of 40 cm in 14 minutes (2.86 cm/min), and the second by a slower rate of 13 cm in 41 minutes

Figure 8.4. Flat-bottomed channel at minute 8.5.

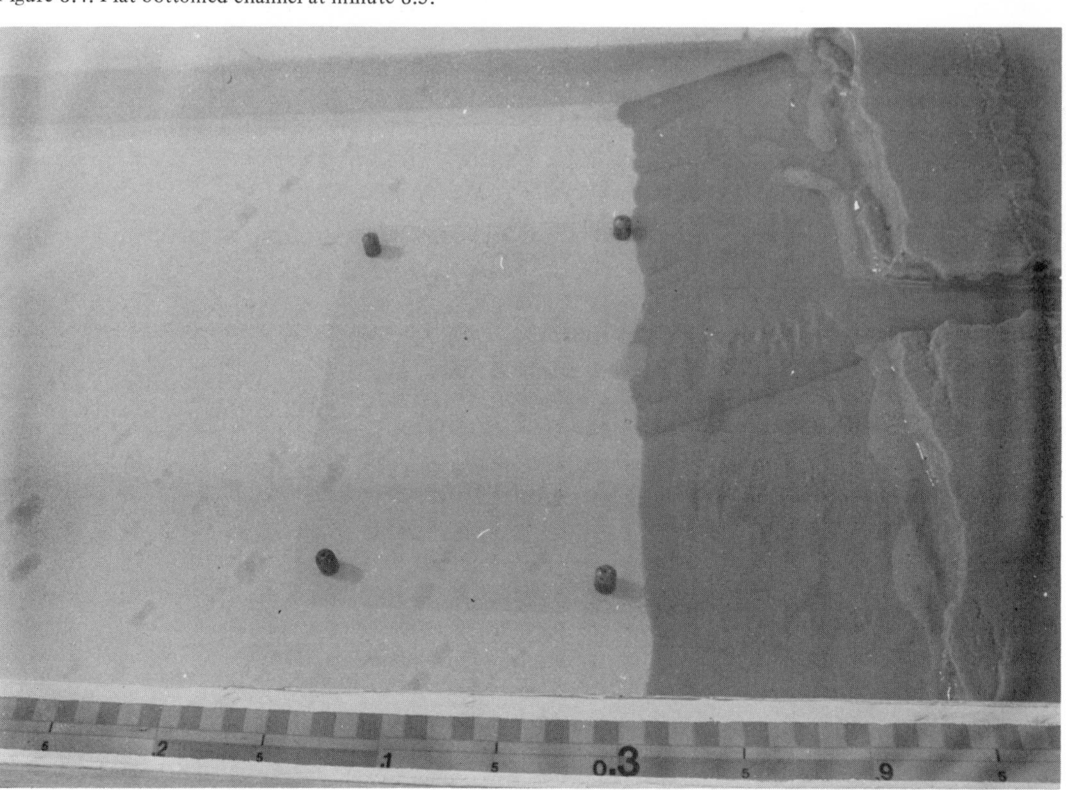

(0.32 cm/min). It may be hypothesized that the first phase represented rapid erosion to a graded profile lying upon the initial water table, while the slower second phase represented a continuing lengthening of the graded segment as the water table adjusted to the presence of a developing stream channel by forming a depressed slope upstream from the nick-point (Figure 8.9).

Figure 8.5. Retreat of nick-point (Experiment 1).

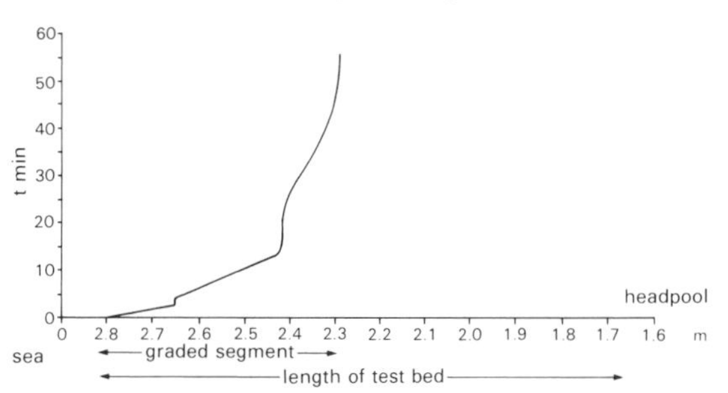

As graded conditions were reached, the energy of flowing water was able to erode laterally to produce a channel (valley) of 4.7 cm maximum width at a position 7.0 cm upstream from the mouth. This width increased from the initial groove width of 3.5 cm and narrowed again to 3.5 cm when overhanging banks closed in after MINUTE 35.

Water depth at the 7.0 cm upstream section remained constant throughout the experiment, and channel (valley) depth remained constant at 2.4 cm from MINUTE 25 to switch-off, indicating that a true graded condition existed at this section after 25 minutes of running time (Figure 8.7).

Upstream from the nick-point, the channel section was essentially a modified artificial groove in which the initial V-section was converted to a trapezoidal section, and the redistributed material was spaced in centre-bars or riffle zones at 30 cm intervals. If channel curvature is related to stream width in the ratio of about 1:7, and the stream

Figure 8.6. Test bed at end of Experiment 1, showing lateral erosion of channel.

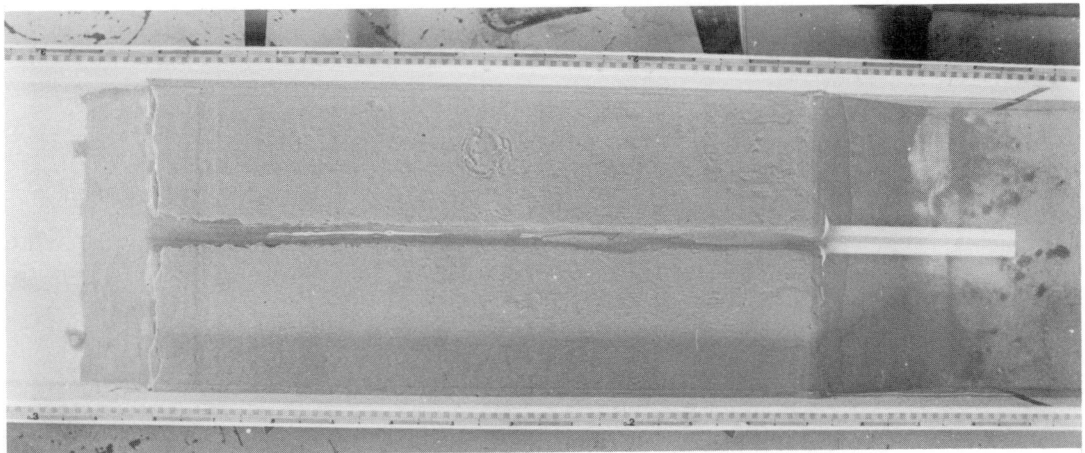

Figure 8.7. Width of valley and depth of channel in Experiment 1, monitored at 7 cm from mouth of channel.

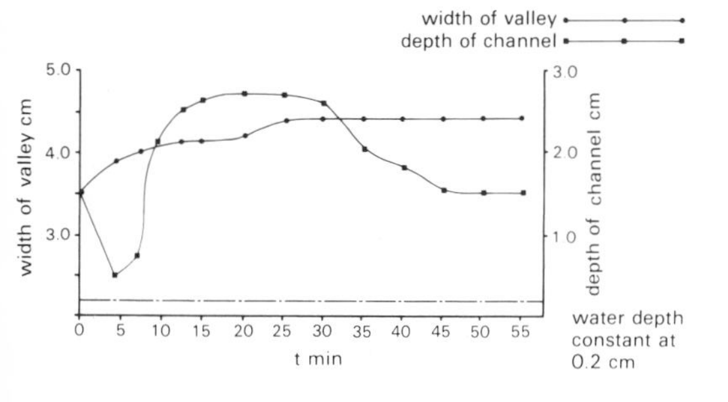

width in the graded condition is about 4.0 cm, this supports empirical evidence from other sources.

Upstream from the nick-point, flow was partly turbulent and partly laminar and sand particles moved initially by low-trajectory saltation or bed traction until the trapezoidal section had developed, after which sediment transport ceased.

Downstream from the nick-point, turbulent flow existed most of the time, and material moved by slightly higher-trajectory saltation and bed traction. It was dislodged from the base of the banks and from the bed, the resulting bed grooves shifting across the section. Banks collapsed by undercutting, caused apparently by stream flow, but the potential existed for effluent seepage from the groundwater body, and banks weakened by basal sapping folded in towards the stream centre before fracturing and falling in. The material dumped into the stream was removed by stream flow until a condition was achieved resembling that immediately prior to the dumping, demon-

Figure 8.8. Deposition in delta at end of Experiment 1.

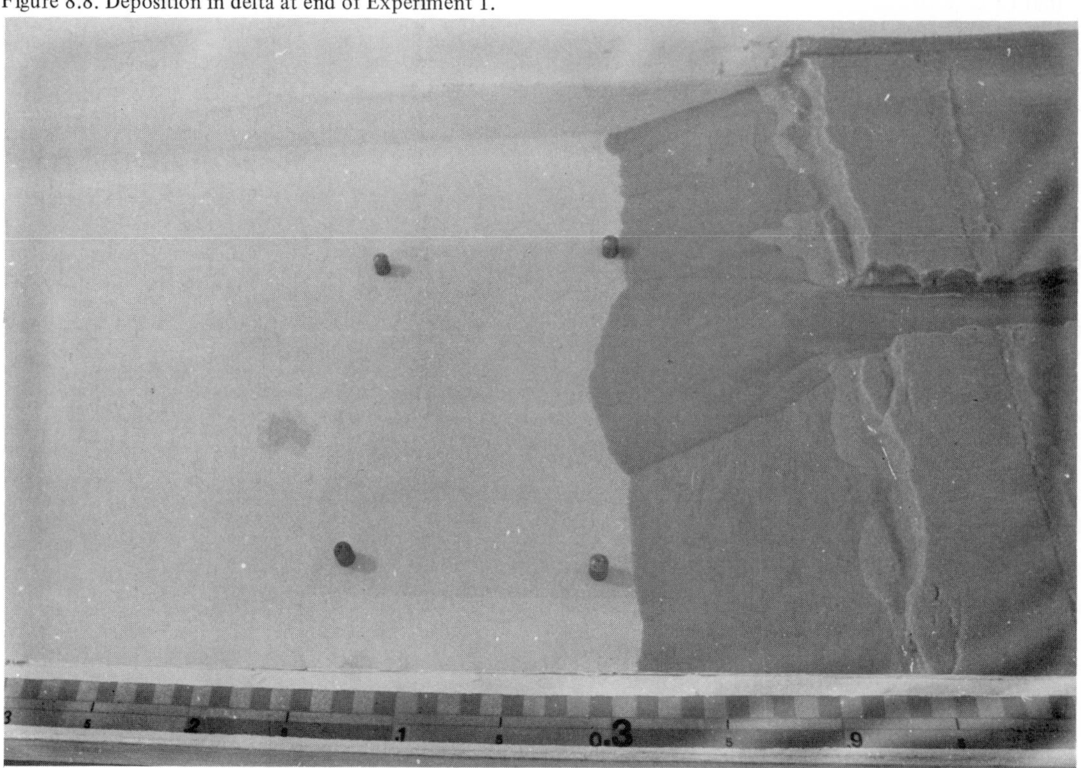

Figure 8.9. Changing water-table and channel profile with lowering sea-level.

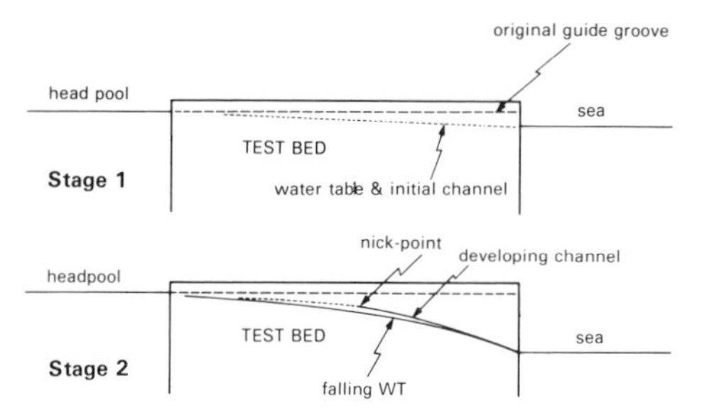

strating a strong hydraulic control over channel geometry.

As the delta grew out, its upper surface was about 2.0 mm beneath sea level - indicating that the stream was effective as an erosive agent beyond its own mouth and across the surface of the delta. *Systems interpretation:* The morphological subsystem is the stream channel from headpool to sea, and at least two different reaches could be determined during the period of the run. Downstream the channel form was graded to base-level (water table), while upchannel the form was essentially that of the original guide groove, with a redistribution of material from the V-slopes to form a flat-bottomed section.

The cascading subsystem comprises the input of water into (and from) the headpool which, at 20 cm s^{-1}, gives a rough mass input in 55 minutes of 66 000 g of water plus 309.8 g of sand. Since the fall in height from headpool to sea-level is $10.9-9.3=1.6$ cm, the energy of water through the system is roughly (mass x fall)/10 197.16 = $66\,000 \times 1.6/10\,197.16 = 10.36$ J.

This rough value will be affected by:
(1) evaporation from headpool;
(2) evaporation from sand surface;
(3) evaporation from stream surface;
(4) base flow between headpool and sea, and stream and sea.

(1) Evaporation from water surfaces was estimated by floating a petri-dish of water in the sea during the run, detecting a loss of 7.31×10^{-5} g cm^{-2} min^{-1}, which represents a headpool loss in 55 minutes of 36.5 g.

(2) Evaporation from the saturated sand surface was estimated by leaving a petri-dish of saturated sand in the test surface, detecting a loss of 1.1×10^{-4} g cm^{-2} min^{-1}, which represents a sand surface loss of 37 g.

(3) Evaporation from stream surface was taken to be of the same order as that from the sea or headpool and was therefore 1.24 g during the run.

(4) Influent seepage from stream to groundwater would be an output from the primary system, while that from headpool to groundwater would be a loss of input to the primary system. In practice, the rate of flow through the medium sand, with a hydraulic head of 1.6 cm and a length of 117 cm, would be very slow and is ignored.

The input to the stream is therefore 66 000 g of water minus 36.5 g (evaporation from headpool) minus 37 g (evaporation from sand which will be replenished mainly from the headpool) = 65 926.5 g, with 10.34 J of energy. If the loss from the stream surface is taken as 1.24 g but only half the energy of this water is lost (because water evaporating from downstream will already have performed work), then energy loss is $(1.24 \times 1.6)/(10\,197.16 \times 2) = 9.73 \times 10^{-5}$ J, thus giving a mass output of $65\,926.5 - 1.24 = 65\,925.26$ g of water with energy of $10.34 - 9.73 \times 10^{-5} = 10.3399$ J, plus the sand mass in the delta and its kinetic energy.

Clearly, the losses from evaporation are all of a lower order of magnitude and in a quantitative sense may be ignored.

The sand enters the system by erosion from within, but is none-the-less an input with potential energy.[8] The load mass is assessed from deltaic deposits (309.8 g in this experiment) and its potential energy will depend upon the height of its origin in the stream above sea-level. Little or none will originate upstream from the nick-point (53 cm from mouth at switch-off), but the upper edge of the nick-point will be at about headpool level. Hence total drop in elevation, H, is $10.9-9.3=1.6$ cm.

But it must be assumed that stream load derived from all points downstream from the nick-point, so that a mean height of origin should be taken and, assuming again that the downstream profile is linear, $\bar{H}=\frac{1}{2}H=0.8$ cm. Thus potential energy of the load is (mass x height)/10 197.16 = $(309.8 \times 0.8)/10\,197.16 = 0.024$ J.

Total input is therefore water mass + sediment mass + water energy (10.34 J) + sediment energy (0.024 J). Output from the stream is water mass + sand mass + (10.34 J + 0.024 J − loss of mechanical energy in work done and friction) + heat energy (Figure 8.10).

During the run the morphological subsystem was modified. The initial artificial guide channel rapidly changed by incision to base-level to approach the graded condition by MINUTE 14

(Figure 8.5), but channel deepening did not cease downstream until MINUTE 25, when true equilibrium of channel depth was achieved in the downstream section. Downstream channel width tended to increase for the first 14 minutes and then to remain constant (except for the infolding effect). Upstream, an equilibrium condition obtained from the commencement of flow to switch-off, as the stream did not become graded and true channel widening did not occur. The nick-point cut back 53 cm from the mouth by switch-off leaving an upper segment of 64 cm in unstable equilibrium, since, owing to lack of time, it had not adapted hydraulically to the total geomorphic control factors.

An example of a negative-feedback loop existed during periods of bank collapse. The material falling into the stream decreased its width, increased its depth, slope and stream velocity, and rapidly accelerated erosion of the debris. The erosion served to increase channel width, decreasing channel depth, slope and stream velocity, effectively restoring the original state of dynamic equilibrium (Figure 8.11).

Comparison of normal and systems appraisals: The appraisal of process and form in the normal way permits more flexibility in presentation of general experimental cause-and-effect relationships. For instance, the focus can move from total bed to channel, and from channel to pool and riffle features with little loss of clarity.

However, the main theme is channel development, and systems interpretation requires specific definition of the channel and the mass and energy directly associated with it. The sand-bed–groundwater body is clearly seen to be a separate system although it governs aspects of the channel process-response system. Similarly, the delta is seen to be a separate system, although containing output from the channel system. In thus separating the various systems, their relationship to each other becomes more evident, and the attributes that could themselves be quantified clearly emerge, indicating other possible experiments, in this case groundwater, delta or pool and riffle development.

Figure 8.10. System diagram: graded and ungraded channel segments (Experiment 1).

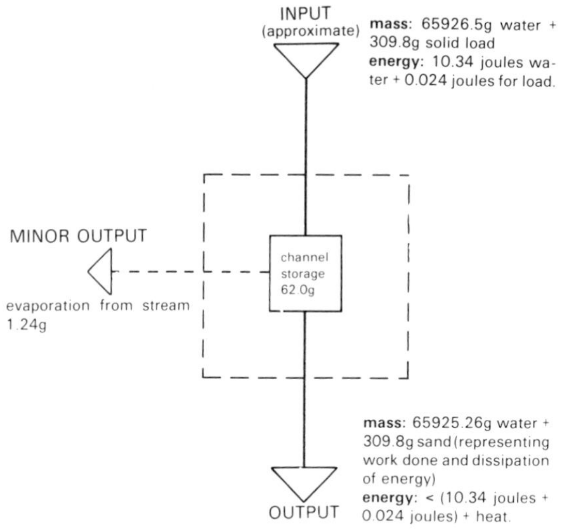

Experiment 2

Purpose: To develop stream terraces by changes in base-level and discharge.

Apparatus: Flume with adjustable delivery and variable sea-level; medium sand formed into a test bed 1.08 m long; sliding carriage with straight blade and channel scribe; brush, measuring cylinder and other glassware; timer, balance.

Procedure: Treatment of sand and development of bed was as in Experiment 1, to produce a horizontal bed with a single central groove as in Figure 8.12. Sea-level, headpool and discharge were set for phase 1 of the run and deltaic material was collected at end of the phase. Discharge and sea-level were reduced for phase 2 and again for phase 3 and deltaic material was again collected and weighed after each phase. A section at 30 cm

Figure 8.11. Negative-feedback loop (Experiment 1).

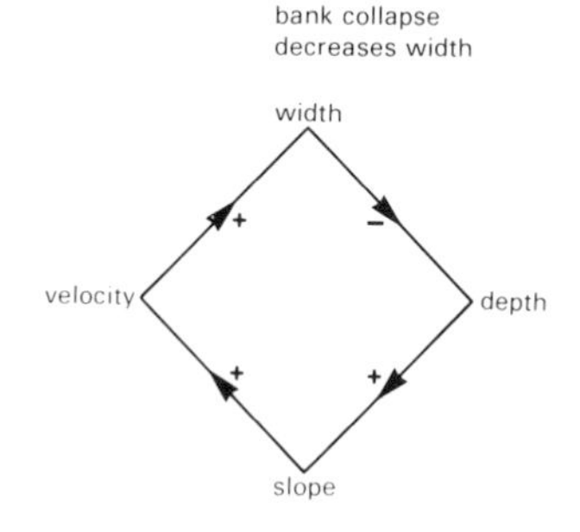

upstream from the mouth was used to monitor channel and valley form changes.
Controls: (Scale 1:1) *Phase 1* - discharge $39.15\ cm^3 s^{-1}$; sea-level constant at 9.5 cm depth; headpool constant at 10.9 cm depth; duration of phase 20 minutes. *Phase 2* - discharge set at $19.72\ cm^3 s^{-1}$; sea-level constant at 8.4 cm depth; headpool constant at 10.6 cm depth; duration of phase 20 minutes. *Phase 3* - discharge $10.37\ cm^3 s^{-1}$; sea-level constant at 7.4 cm depth; headpool constant at 10.3 cm depth; duration of phase 20 minutes.
Observations and data: (*phase 1*) (discharge, $Q = 39.15\ cm^3 s^{-1}$). Turbulent flow produced erosion and transformation of the total channel length to rectangular section. Bank undercutting occurred and a tendency developed for the channel to be narrower than the valley in the downstream segment (Figure 8.13 at MINUTE 19). A post-phase-1 photo (Figure 8.14) shows the final degree of erosion and deltaic deposition after 20 minutes running time. The monitored section at 30 cm upstream from the mouth gave a valley floor width of 6.0 cm and a depth from initial surface to stream bed of 1.8 cm. Stream channel depth was 0.3 cm. The existence of a stream channel narrower than the valley floor was not clearly perceptible at this section, but at the mouth the channel was 4.4 cm in width and 0.3 cm in depth while the containing valley was 7.0 cm in width.
Observations and data: (*phase 2*) ($Q = 19.72\ cm^3 s^{-1}$). The stream immediately began to incise to the lowered base-level leaving a terrace in the downstream section. Upstream the lesser discharge filled the channel floor (Figure 8.15). At MINUTE 20, at the 30 cm section the total valley width was 6.0 cm, the total valley depth (to stream bed) was 2.75 cm and the channel width was 4.4 cm giving at this location $6.0 - 4.4 = 1.6$ cm of older valley floor as terrace. Channel depth was 0.2 cm. At the end of the 20 minute phase the channel mouth (as distinct from the valley) was 5.0 cm wide and 0.2 cm deep. Terraces had been eroded to some extent.

Figure 8.12. Test bed for Experiment 2.

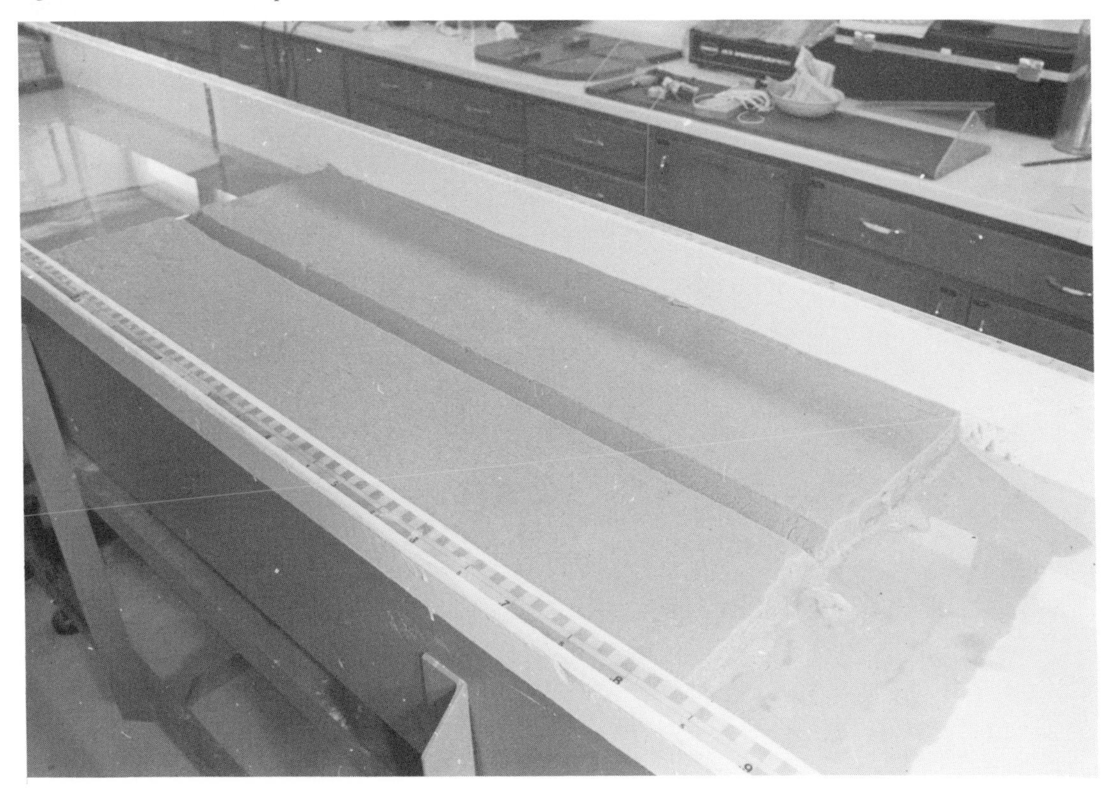

Figure 8.13. Test bed at minute 19 of phase 1 of Experiment 2.

Figure 8.14. Experiment 2 at end of phase 1.

Observations and data: (phase 3) ($Q = 10.37$ cm^3 s^{-1}). Re-incision commenced immediately at the mouth, with a narrower stream channel and development of a lower terrace (Figure 8.16). The channel at the 30 cm section was 2.5 cm wide and its bed was 2.85 cm below datum and 0.1 cm deep. Valley width at the top remained at 6.0 cm. At the mouth the channel flared to 8.0 cm in a 7.3 cm wide and 4.0 cm deep (to stream bed) valley. At the 30 cm section the overall valley-channel section is as indicated in Figure 8.17.

The oven-dried material collected from the deltas at the end of each phase weighed: phase 1, 522.5 g; phase 2, 252.2 g; phase 3, 227.3 g.
Conclusions: The reduction of discharge in a stream is capable of producing terraces by channel-shrinking if the ability exists to downgrade towards the water table (or base-level). In the three phases of the stream's existence approximate grading was achieved within a few minutes of commencement (see Figure 8.15 at MINUTE 8.5 of phase 2). The lowering of base-level between phases permitted rejuvenation; each successive discharge was of increased effectiveness in grading upstream, but in the furthest upstream segment the stream bed remained unchanged after phase 1, so that terracing is only to be seen in the downstream portion. The oldest, highest terrace, which retains a terracette (Figure 8.18) from the oldest channel/floodplain association, dies out upstream at 38 cm from the mouth (2.36 m on flume edge), while the later phase 2 terrace dies out at 60 cm from the mouth (2.15 m on flume edge) as indicated in Figure 8.19.

Traces of terracettes, left behind in floodplains as the channel at various times shifted across the valley floor, are to be seen at all levels in Figure 8.20. There are therefore two distinct processes at work in this experiment, the first related to change of discharge and rejuvenation, and the second related to the shifts of a channel as it approaches a graded condition. The drop in sea-level was about 1.0 cm between phases, but the headpool remained at more or less its original level, so erosion should be related to energy, but for simplicity the relationship of erosion to discharge may be assessed by

Figure 8.15. Experiment 2 at minute 8.5 of phase 2, showing incision downstream but discharge filling old channel upstream.

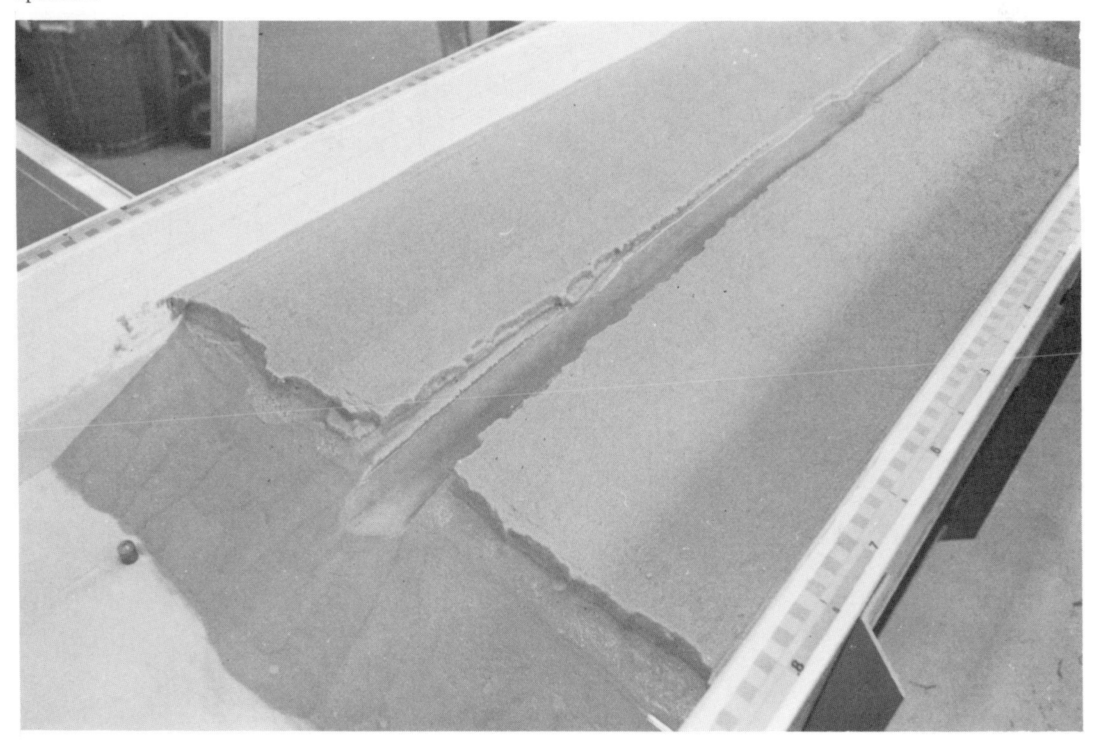

reference to deltaic material as shown in Figure 8.21.

Systems interpretation: In effect, there are three systems occupying in time sequence the same spatial position. In so far as the stream flow was continuous, in a non-laboratory example the phenomenon may be considered as a single system twice perturbed to give three phases - which was the stance taken above in developing conclusions.

It is necessary to recognize, however, three magnitudes of input, throughput and output, with resultant modified morphological subsystems, involving three phases of grading and approach to dynamic equilibrium.

The encompassing system is not, as in Experiment 1, simply the channel, since part of the exercise is to relate terraces to stream action, and this incorporates valley features. It seems advisable, therefore, to envisage three morphological stream subsystems within a valley system, each with its appropriate cascading subsystem (Figure 8.22).

By the end of the initial phase a valley was produced 108 cm long, 6.0 cm wide (at 30 cm

Figure 8.16. Experiment 2 phase 2 at minute 20, showing some terrace at this stage.

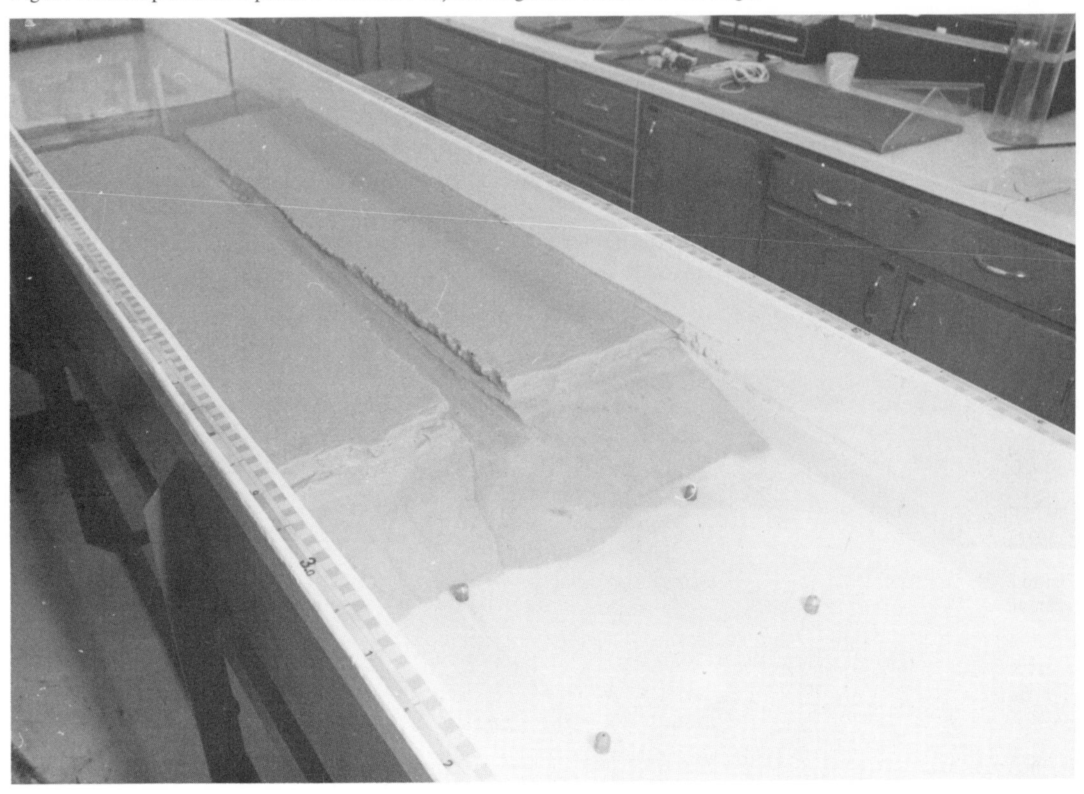

Figure 8.17. Overall valley-channel section at end of phase 3 of Experiment 2, taken at 30 cm upstream from the mouth.

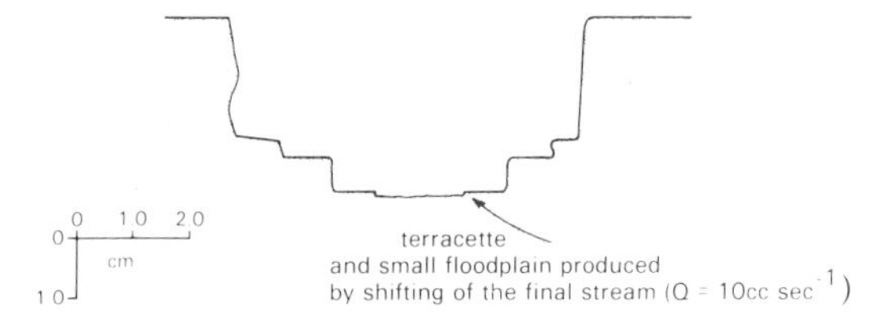

upstream) and 1.8 cm deep, containing in its floor a stream about 4.0 cm wide and 0.3 cm deep. The input to this initial system was 48 000 g of water with 48 000 × 1.4 (drop in height from headpool to sea-level)/10 197.16 = 6.6 J potential energy plus 522 g of sand (from erosion) with 0.036 J potential energy. This passed through the system expending energy in mechanical work in eroding and transporting sand, so that the output was 48 000 g of water + 522 g of sand with <6.636 J

Figure 8.18. Test bed at end of Experiment 2, showing fade-out of terraces upstream (oblique view).

Figure 8.19. Test bed at end of Experiment 2, showing fade-out of terraces at 38 cm and 60 cm from the mouth (plan view).

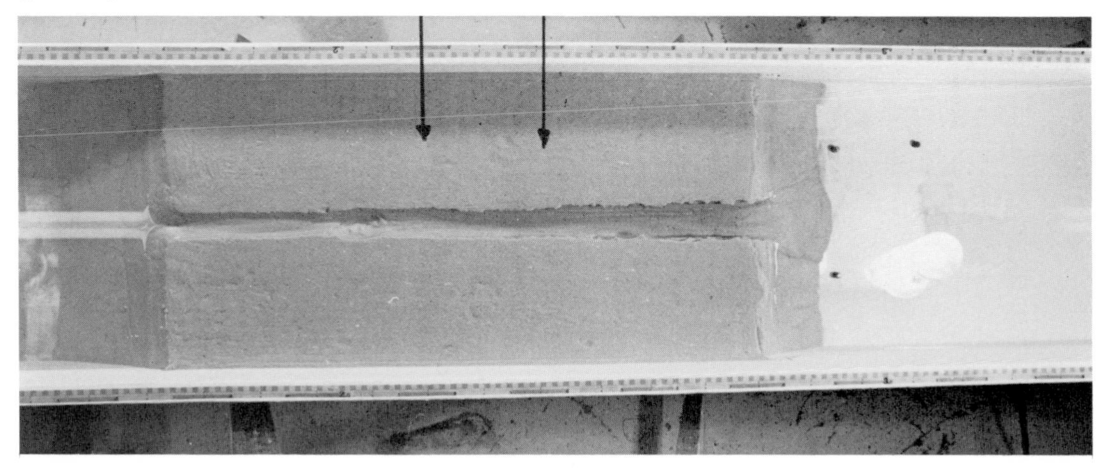

kinetic energy (E_k) + heat energy. The loss by evaporation will be ignored (see Experiment 1).

The morphological system of valley and stream also included, at the end of the phase, terracettes caused by shift of the stream across the valley floor, and achievement of a condition of dynamic equilibrium may be inferred.

Phase 2 commences with a reduction of mass and energy input to a value which will sum to 24 000 g of water with (24 000 × 2.2)/10 197.16 = 5.2 J potential energy (E_p), and 252 g of sediment with 0.01 J (E_p).

This value of 0.01 J is derived in the following way. The profile inherited by phase 2 from

Figure 8.20. Experiment 2. Post-run view of total stream length showing relationship of older channels to final channel and terrace flight.

Figure 8.21. Decrease in delta mass with decrease in energy (Experiment 2).

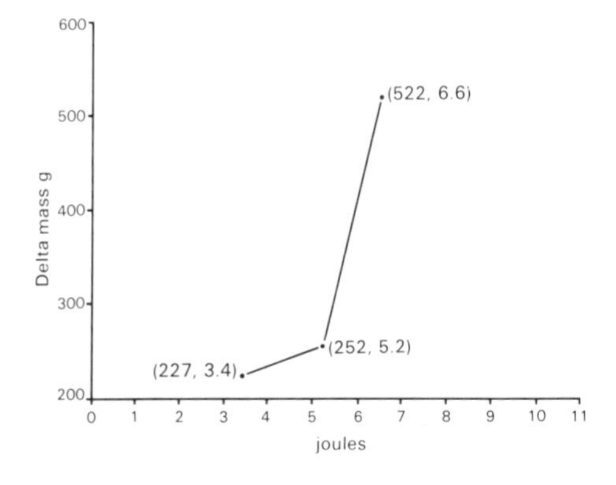

phase 1 is 108 cm long and slopes down from headpool to sea-level, but only 38 cm of this is actively eroded in phase 2. Thus the height of origin of sand particles can be no higher than the height of the channel 38 cm upstream. Assuming the profile inherited from phase 1 is linear, then H in phase 2 is $(2.2 \times 38)/108 = 0.77$ cm, and only half of this is to be taken $(\bar{H}) = 0.39$ cm. Therefore E_p is $(252 \times 0.39)/10\,197.16 = 0.01$ J.

Reduction of Q and lowering of base-level perturbs the system, and while the channel cross-section diminishes, the stream renews vertical erosion.

As the modified throughput works towards producing a new condition of equilibrium (grade) the valley deepens along a restricted floor width introducing terraces to the morphological system. The new base-level permits the stream to incise further up-valley than in phase 1, so that the terrace feature ends downstream from the new nick-point. As a new equilibrium is achieved within 20 minutes of phase 2 (note floodplain terracettes) relaxation time is less than 20 minutes. The output from the system during this phase is the 24 000 g of water and 252 g of sand and an unmeasured but small quantity of heat, together with something less than 5.21 J kinetic energy (E_k).

Phase 3 repeats the perturbation and response of phase 2, with input reduced to a total of 12 000 g of water with energy of $(12\,000 \times 2.9)/10\,197.16 = 3.4$ J, plus 227 g of sediment with 0.018 J E_p. The calculation of E_p for stream load is the same as that for phase 2, except that effective stream length is 60 cm and $\bar{H}$ becomes 0.8 cm. Again the stream is able to incise to a new and steeper water table surface related to the lowered base-level, and it grades quickly to an upstream position beyond that of phase 2. A new terrace is produced extending further upstream than that of the phase 1-2 perturbation. Again floodplain terracettes testify

Figure 8.22. Systems diagram relating valley, channel and terrace morphological subsystems (Experiment 2).

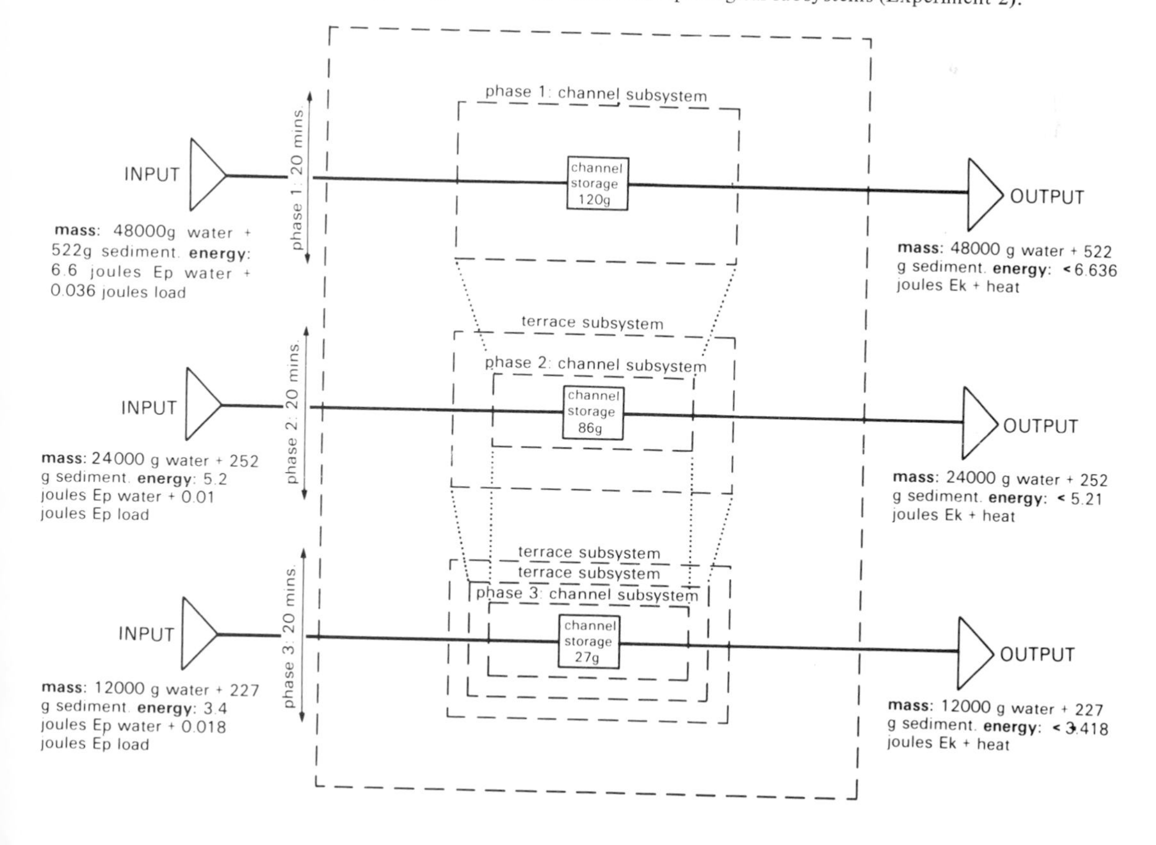

to lateral shift of the stream across the latest graded surface, and therefore to an equilibrium condition in the vertical plane. Output is 12 000 g of water and 227 g of sand, with somewhat less than 3.418 J E_k plus heat energy.

It is interesting to note that in phase 3, E_p for stream load of 227 g is greater than phase 2 E_p for stream load of 252 g (i.e. 0.018 and 0.01 J). This results from a greater $\bar{H}$ value in phase 3 due to stream incising further upchannel. This again reflects the effect of lowering base-level of erosion, which in this case overcomes a lesser Q-value.

The final process-response system includes a complex morphological system with total valley, terraces, terracettes and stream channel. It was formed by a grand total input of 84 000 g of water and 15.26 J of potential energy, with an output of 84 000 g of water, 1001 g of eroded sand, somewhat less than 15.26 J of potential energy and an unmeasured quantity of heat produced by friction.

It may be assumed that during any of the three phases, in which controls were constant, once grading was achieved, momentary perturbation was only in the form of collapsing material (more particularly in phase 1 while the major valley was forming) and that negative feedback rapidly restored the channel system to a state of dynamic equilibrium (see the discussion in Experiment 1).

Experiment 3

Purpose: To produce a floodplain and to examine discharge and linear parameter relationships.
Apparatus: Flume with adjustable input and sea-level control; medium sand; former; glassware; timer.
Procedure: A test bed of medium sand was set up in the flume in such a position as to leave a large headpool: the head of the bed was at 1.56 m, the foot of the bed was at 2.74 m giving a bed length of 1.18 m. The bed was given a plane horizontal surface and a V-groove was emplaced with a scriber attached to the carriage blade. The model was trimmed and brushed and water and headpool water were carefully introduced, each to the bottom of the V-groove. The water supply was set at a low value and the experiment was run for sufficient time to allow channel equilibrium for the initial discharge. Then the discharge was increased five-fold to produce flood conditions and a floodplain. At the cessation of flood discharge, sea-level was lowered slightly to permit regrading, and the initial low discharge was re-employed. Note: in order to use conveniently two Q-values in one experiment, two feed pipes are required, the lower-discharge pipe being plugged with a bung while the larger is open, and vice versa.
Controls: (Scale 1:1) *Pre-flood phase* - $Q = 19.29$ cm^{3}s^{-1}; headpool constant at 10.5 cm depth; sea-level constant at 9.7 cm; duration 25 minutes. *Flood phase* - $Q = 97.84$ cm^{3}s^{-1}; headpool constant at 10.5 cm depth; sea-level constant at 9.7 cm depth; duration 5.6 minutes. *Post-flood phase* - $Q = 19.29$ cm^{3}s^{-1}; headpool constant at 10.5 cm depth; sea-level constant at 9.3 cm; duration 30 minutes.
Observations and data (pre-flood phase): Flow was turbulent, and owing to the high sea-level just lapping the bottom of the V-groove there was little opportunity for the stream to incise, so that it rapidly produced a channel hydraulically adapted to the discharge, with some flare at the mouth, and pool and riffle features upstream (Figure 8.23). Some bank slumping occurred and incipient terracing resulted from minor channel shift.

The channel/valley parameters at the end of the phase (measured at 14 cm upstream) were valley width 5.2 cm, valley depth to channel bottom 2.1 cm, and the channel lay in the bottom of the valley filling it more or less from wall to wall. Deltaic material resulting from this pre-flood phase was 265 g dry sand (173 cm^3 wet compact sand). Equilibrium conditions were more or less obtained.
Observations and data (flood phase): The increased discharge produced immediate channel widening and upstream overflow onto the initial land surface (Figure 8.24). The channel/valley parameters at the section 14 cm upstream after MINUTE 5.6 were valley width 12.7 cm, and the stream channel coincident with the valley bottom at about 2.0 cm below the land surface. Some curvature of channel trace occurred (Figure 8.24). Deltaic material resulting from 5.6 minutes of flood discharge had a wet, compact volume of 630 cm^3, and a mass of 946 g. Lateral equilibrium did not develop.

Figure 8.23. Experiment 3. Channel development during pre-flood phase.

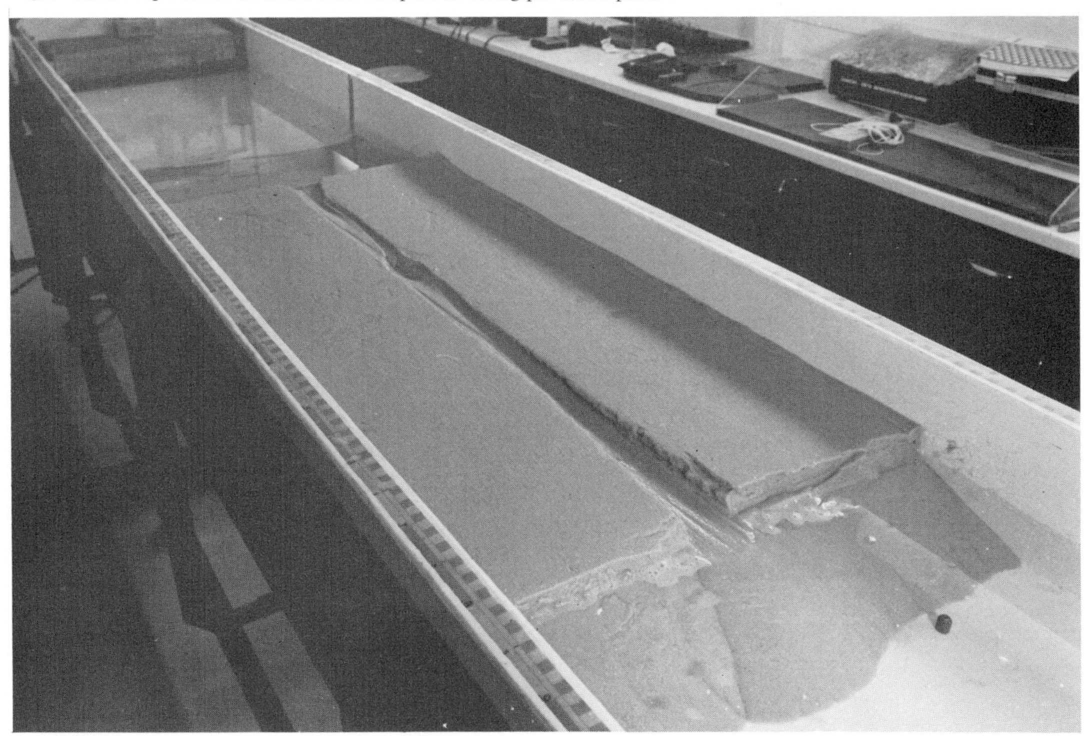

Figure 8.24. Experiment 3. Channel widening and curvature during flood phase.

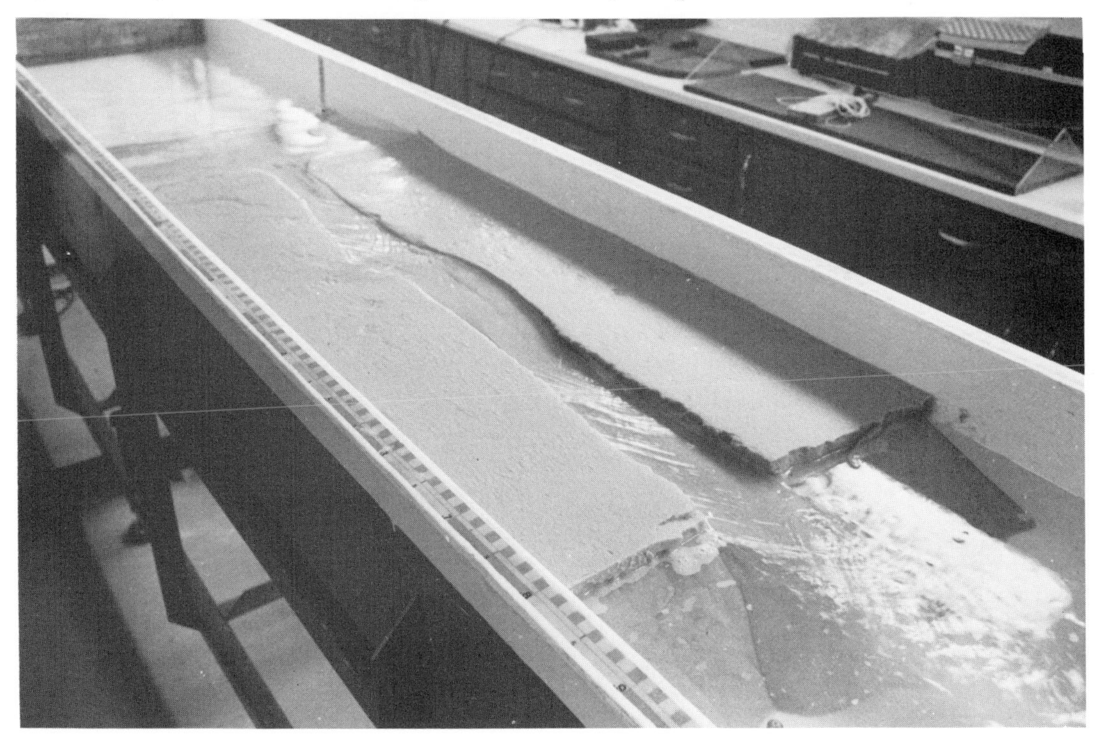

Observations and data (post-flood phase): With discharge back to the normal value the stream shrank in size but did not attain its pre-flood form. With base-level 0.4 cm lower at the mouth some incision was possible, permitting a channel to become defined. The channel took up a curved trace upon the floodplain left by the high discharge. Extensive but irregular floodplain flats occurred in abandoned parts of bends (Figures 8.25 and 8.26). Flow was insufficient to permit the stream to erode laterally into, for instance, the right bank at metre 2.2, although it took up a position hard against the left bank at metre 2.45 to deflect sharply to the right before disgorging into the sea. The channel/valley parameters at the 14 cm section after the run were; valley width 12.7 cm, valley depth (to floodplain) 2.0–1.8 cm, channel width 5.6 cm, channel depth 0.1–0.2 cm.

Conclusions: The flood discharge represented the first flood to occur in the stream segment and was responsible for widening the valley (channel) from 5.2 to 12.7 cm at the monitored section, but did not deepen the channel because of the fixed base-level to which the initial stream had already adjusted. The experiment therefore demonstrates the production of what later becomes a floodplain. The high flood discharge which might in a natural stream occur only once per year, produced its own trace, different from the straight trace of the initial stream.

When, after 5.6 minutes, the discharge returned to a lower value, the stream did not assume its original form. It was a little wider (5.6 cm rather than 5.2 cm) and it curved across the floodplain zone, but not with the same radius of curvature as the flood stream. The post-flood phase ran for 30 minutes. It did not reach an equilibrium condition in this time but established only a new stable channel form. The experiment could be extended to include a second or third flood phase to test for stability of this post-flood 20 $cm^3 s^{-1}$ channel.

Systems interpretation: The lower-value discharge is throughput for a channel morphological subsystem, while the high-value discharge is throughput for a channel plus floodplain morphological subsystem. The overall process-response system is therefore complex with spatial and temporal variation. Ignoring evaporation and base-flow losses, the input was normal flow of 19.29 $cm^3 s^{-1}$ or, during the first phase, $19.29 \times 25 \times 60 = 28\,935$ g of water plus $(28\,935 \times H)/10\,197.16 = 2.27$ J E_p, plus $(0.4 \times 265)/10\,197.16 = 0.01$ J stream load E_p for 25 minutes flow. Output was 28 935 g water + <2.28 J E_k + heat energy + 265 g of sediment. Channel storage during this phase was $L \times W \times D = 118 \times 5.2 \times 0.15 = 92\ cm^3 = 92$ g of water. (H = hydraulic head = 0.8 cm and $\frac{1}{2}H$ is used for load energy.)

During the flood phase water flows from the supply zone into the channel and onto the floodplain at the rate of 97.84 $cm^3 s^{-1}$ for 5.6 minutes, giving an input of $97.84 \times 5.6 \times 60 = 32\,874$ g of water with $(32\,874 \times 0.8)/10\,197.16 = 2.58$ J E_p plus 0.04 J E_k for load. Storage within the floodplain plus channel subsystem is about $118 \times 12.7 \times 0.15 = 225\ cm^3$ (225 g). Output is 32 874 g of water plus 946 g of sediment plus <2.62 J E_k + heat.

These subsystems may be represented canonically as in Figure 8.27. During normal flow, input is only to the channel subsystem which adapts geometrically to constant hydraulic conditions. At the onset of flood input, the floodplain is formed (or reoccupied) and the channel proper is distorted or subjected to perturbation. In the present case the original channel was considerably changed. When normal flow recurred, the floodplain was abandoned and a new channel developed within a few minutes (relaxation time), but now curved and somewhat wider than the original. The stream gradient was increased for phase 3 so that the cascading subsystems of phases 1 and 3 are dissimilar. For the purpose of quantitatively relating channel to floodplain only phases 1 and 2 have therefore been used. In phase 3 the re-establishment of a channel adapted to a 19.29 $cm^3 s^{-1}$ discharge represents a negative-feedback loop that is overcoming the effect of both a temporarily increased discharge and a lowering of base-level.

Experiment 4

Purpose: To produce a misfit stream. This experiment is similar to Experiment 2, but a larger initial channel is used to emphasize the misfit phenomenon.

Figure 8.25. Experiment 3 during post-flood phase, showing extensive floodplain flats.

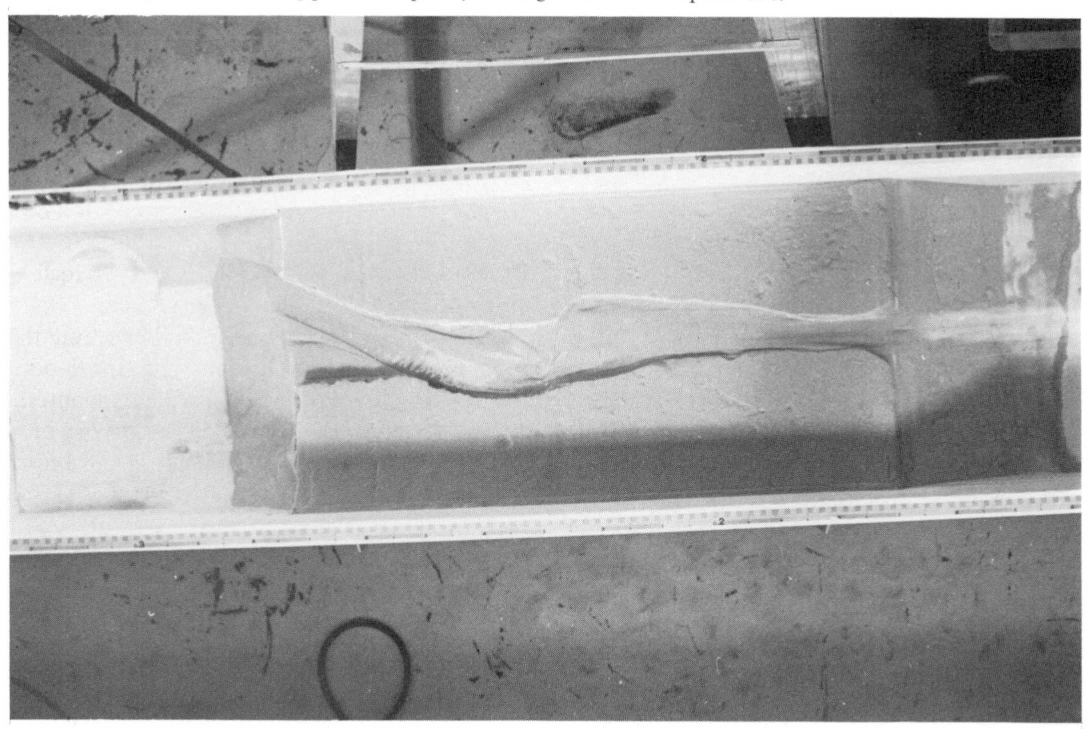

Figure 8.26. Experiment 3 during post-flood phase showing floodplains in abandoned parts of valley floor.

Apparatus: Flume with feed and sea-level controls, medium sand, U-shaped former, glassware, timer, etc.
Procedure: A 1.34 m test bed was formed, being horizontal longitudinally but containing a broad valley 47 cm wide and 3.3 cm deep, with rounded sides. Water was added to the headpool and sea to within 2 mm of the valley floor level, giving very little gradient between head and foot of the intended river system.

The experiment had three phases: *Phase 1* - a high-input phase with constant sea-level to produce hydraulic characteristics in the initial valley (two feed pipes delivering water together); *Phase 2* - a lower-input phase with constant sea-level to produce a lesser stream in the valley (the lesser feed pipe delivering alone); *Phase 3* - as phase 2 but with falling sea-level to enable a misfit stream to incise into the old valley floor.

Controls: (Scale 1:1) *Phase 1* - headpool level 7.25 cm; sea-level constant at 7.0 cm; input 94.19 $cm^3 s^{-1}$; duration 20 minutes. *Phase 2* - headpool level 7.25 cm; sea-level constant at 7.0 cm; input 21.46 $cm^3 s^{-1}$; duration 10 minutes. *Phase 3* - headpool level 7.25 cm; sea-level at start 7.0 cm, dropping at a rate of 0.137 cm/min to a final level of 2.9 cm; input 21.46 $cm^3 s^{-1}$; duration 30 minutes.
Observations and data (phase 1): The initial high input of 94.19 $cm^3 s^{-1}$ flooded the pre-formed valley upstream to a width of 44 cm and produced an inner sinuous thalweg 9.5 cm wide and 0.4 cm deep. Down-valley incision occurred, and at MINUTE 9 the inner thalweg existed as part of a discrete stream flaring slightly to 16.0 cm width (Figure 8.28). At the end of the 20 minute run the mouth had widened to 22.0 cm and was 0.15 cm deep. Dry deltaic sediment weighed 380 g.

Figure 8.27. Experiment 3. Canonical diagrams for normal and flooded channel systems.

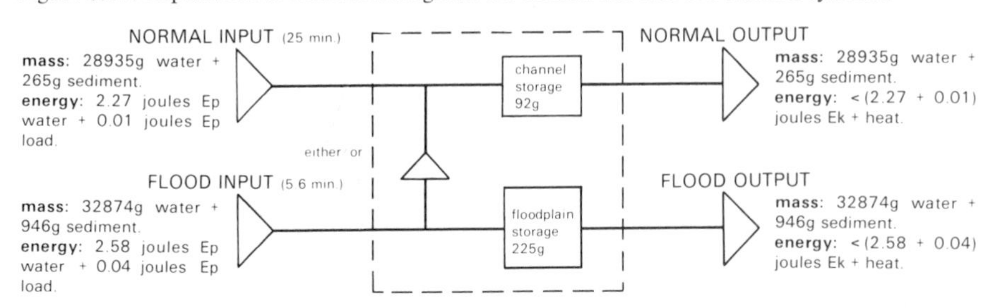

Figure 8.28. Experiment 4 at minute 9 of phase 1.

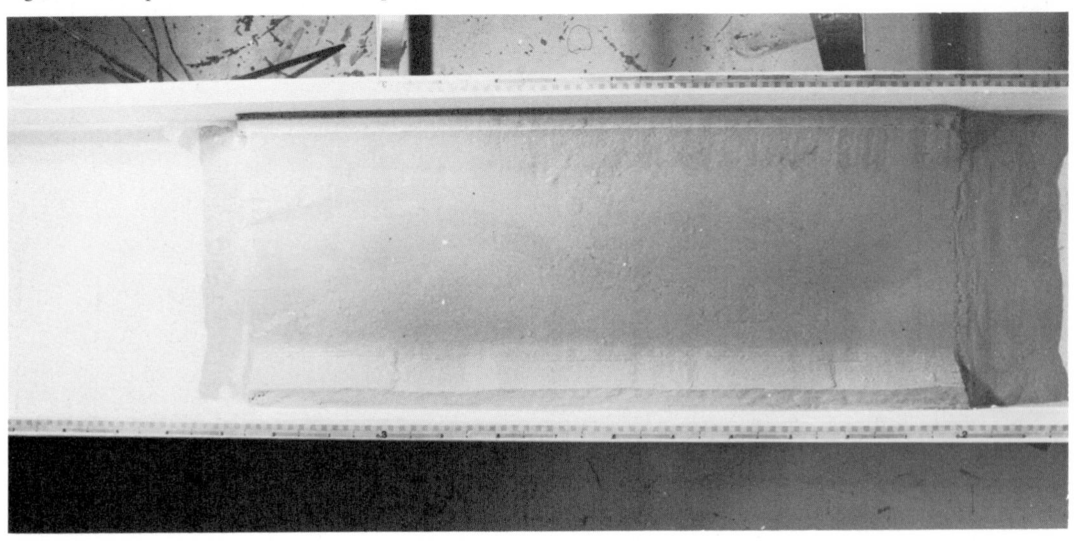

Observations and data (phase 2): With a reduction of discharge to 21.46 $cm^3 s^{-1}$, but headpool and sea-level maintained as in phase 1, there was shrinking of the channel to the previous sinuous inner course with some flooding in upper reaches. At this stage the stream was in a misfit condition and erosion was reduced. Dry weight of eroded sediment was 30.0 g. The result of 10 minutes of stream flow under these conditions was an accentuated sinuous channel with radius of curvature of some 56 cm, compared with a channel width of about 14.5 cm and channel depth of 0.15 cm at mid-valley (Figure 8.29). This curvature was inherited from phase 1.

Observations and data (phase 3): As soon as sea-level began to drop, the misfit stream commenced incision, although retaining areas of floodwater up-valley where the original base-level of phase 1 remained (Figure 8.30). Erosion and sedimentation were pronounced, and a delta rapidly formed. Sea-level fell in 30 minutes from 7.0 to 2.9 cm, and in this time the stream incised up-valley a distance of about 60 cm. At switch-off the rejuvenated section was receiving flow from upstream through a pronounced draw, incorporating a nick-point. The flared mouth of phase 2 was abandoned and traces of lateral surface flow (small rills) occurred on the left bank where phase 2 flow rapidly retracted into the phase 3 channel (Figure 8.31).

Downstream from the nick-point the channel proper narrowed to some 3.7 cm width and 0.23 cm depth in a 13–16 cm valley, formed from the phase 2 channel. At a section 30 cm upstream from the coastal cliff a cross-profile after switch-off showed a stepped or terraced arrangement as in Figure 8.32. The mass of dry deltaic sediment was 1472 g.

Conclusions: The high commencing input of 94.19 $cm^3 s^{-1}$ was sufficient to fill the valley from side to side, and to produce slight slope flexures by capillarity and settling along the valley walls. Flow was mainly laminar but a sinuous inner channel formed immediately as a faint trace, winding from side to side of the valley. With only a slight gradient to energize the flow erosion was moderate, but after 20 minutes the valley had become adapted to the discharge and was essentially a stream channel.

When input diminished to 21 $cm^3 s^{-1}$ the flow in the downchannel segment became restricted

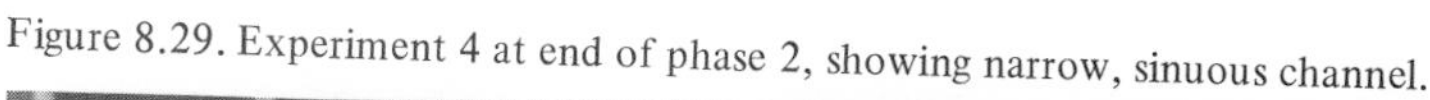

Figure 8.29. Experiment 4 at end of phase 2, showing narrow, sinuous channel.

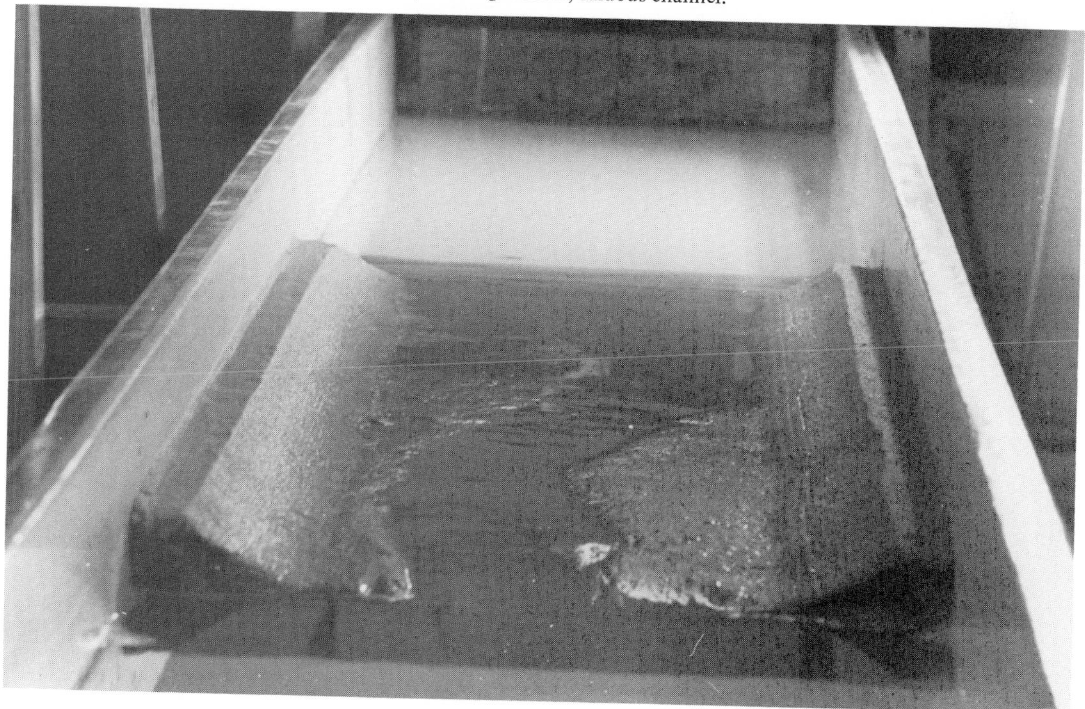

Figure 8.30. Experiment 4 soon after start of phase 3, with incision of misfit stream.

Figure 8.31. Experiment at end of phase 3, showing final relationship of misfit stream to original fluvial valley.

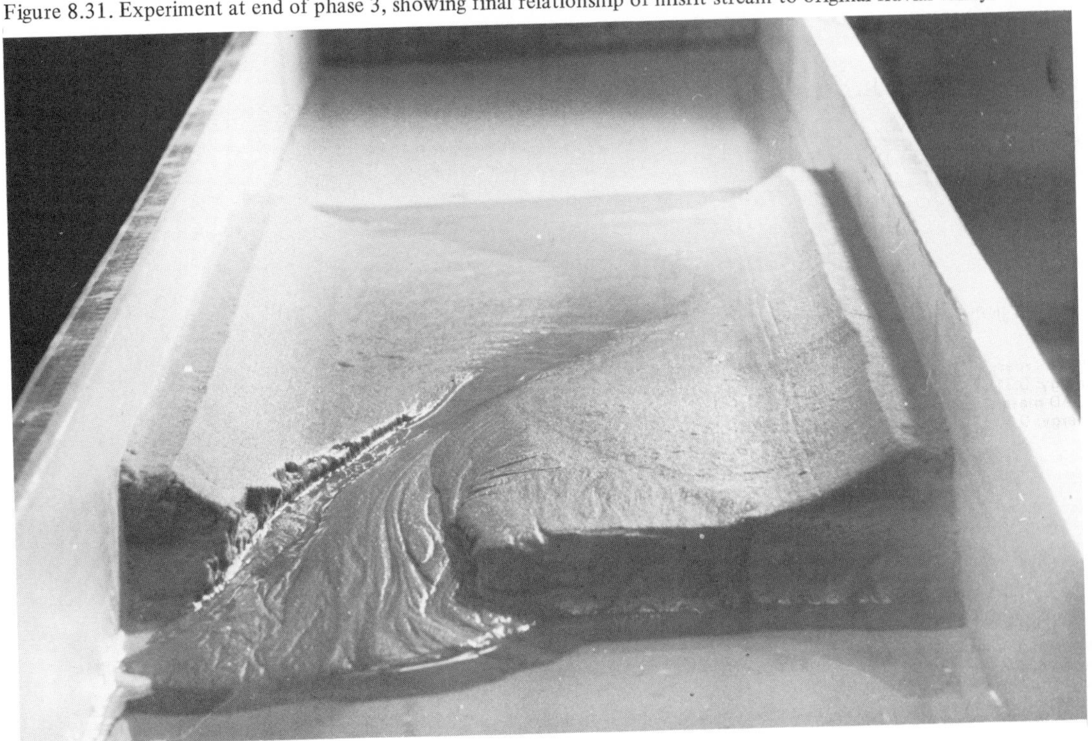

to the sinuous inner channel, and therefore became a misfit stream in relation to the hydraulically adjusted prior channel (valley). Erosion was minimal during this phase due to the greatly reduced energy component.

Upon re-vitalizing the stream by lowering the base-level of erosion, the channel narrowed, incised and then eroded laterally at the mouth, flagellating across a widening segment of the initial channel (now the large containing valley).

Systems interpretation: The three-phase nature of the experiment shares factors with Experiments 2 and 3. The containing morphological subsystem is the total valley/channel form which contains smaller elements associated with different throughputs: firstly, a simple misfit channel, barely able to manifest itself in its morphological environment inherited from phase 1; and finally the rejuvenation of the misfit stream, which involves entrenching to produce a more specific misfit form. The process-response system is depicted in Figure 8.33, in which the time factor is given a place by representing the three morphological subsystems with their throughputs in sequence.

In the 20, 10 and 30 minutes respective duration of each phase equilibrium conditions did not occur, although they were probably close in phase 2. In phase 3, perturbation was strong and with an energy-high system reaction was commensurately strong, but even so a relaxation time of

Figure 8.32. Cross-section of channel at end of Experiment 4, taken at 30 cm upstream from the coastal cliff.

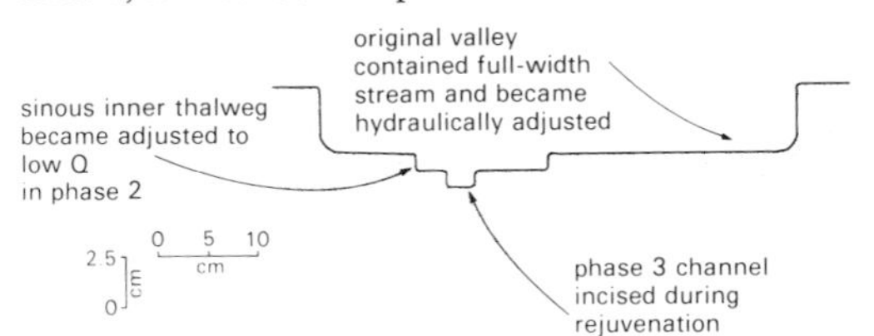

Figure 8.33. Process-response system (including all transient forms) for Experiment 4.

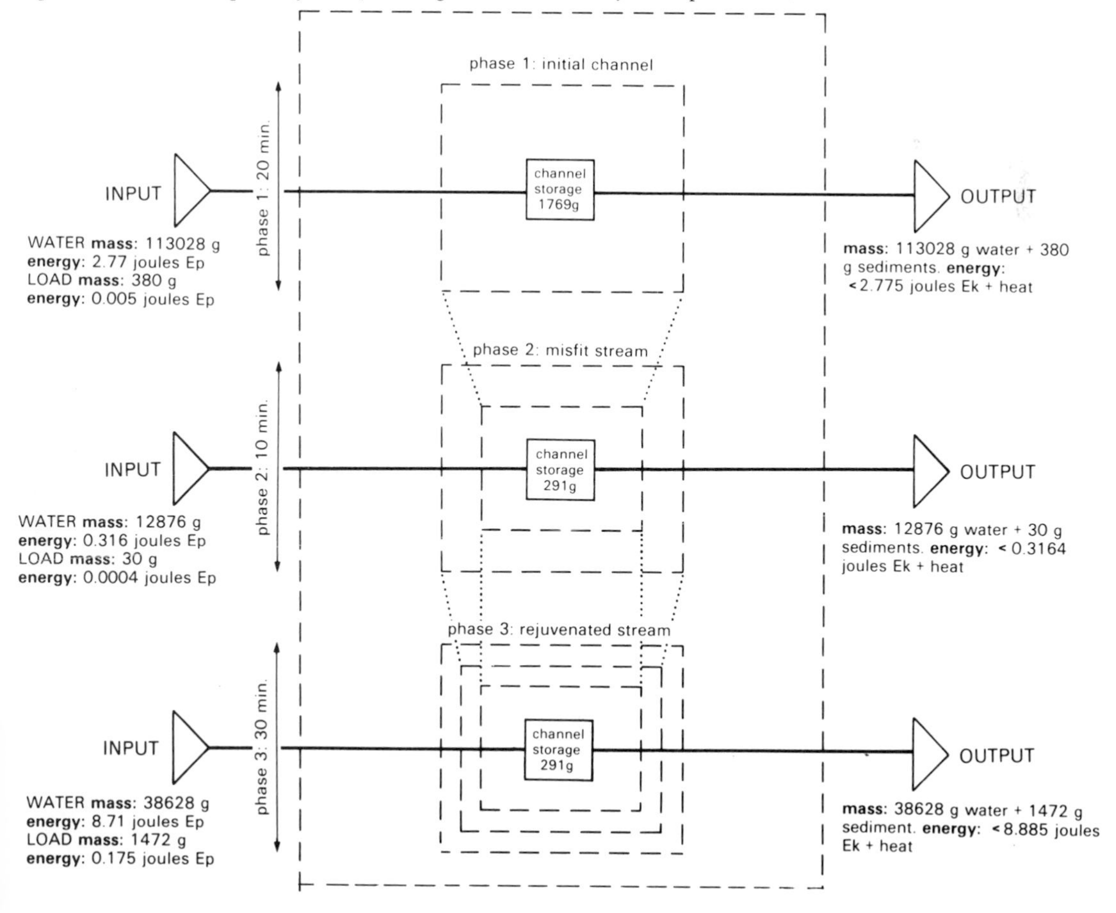

considerably more than 30 minutes would be indicated. In this type of model the ultimate equilibrium form would be a fairly featureless inclined surface, lacking evidence of initial valley (channel) or misfit features. This condition of equifinality (Davisian peneplain) develops in several hours or several days depending upon controls used.

Phase 1: Input - water mass 113 028 g, $E_p = 2.77$ J; sand mass 380 g and E_p derived from $\frac{1}{2}H \times$ mass $= 0.005$ J. Output - mass, ignoring evaporation and base-flow from stream, 113 028 g of water + 380 g of sediment plus less than 2.775 J kinetic energy plus heat energy from friction and mechanical work done.

Phase 2: Input - water mass 12 876 g, $E_p = 0.316$ J; sand mass 30 g, $E_p = 0.0004$ J. Output - mass as input, with kinetic energy less than 0.3164 J, plus heat energy.

Phase 3: Input - water mass 38 628 g, $E_p = 8.71$ J; sand mass 1472 g, $E_p = 0.175$ J. Output - mass as input, with kinetic energy less than 8.885 J, plus heat energy.

Experiment 5

Purpose: To examine the erosion of bends in streams.

Apparatus: Flume; former; medium sand; timer and various glassware.

Procedure: A test bed was formed, 1.16 m in length with a flat, horizontal surface and containing a guide groove of circular section, 1.6 cm at its centre depth and 4.2 cm wide. This groove was scooped out, following a curved trace scribed round the edge of a circular plastic container of 8.25 cm radius to give $6\frac{1}{2}$ bends (Figure 8.34). Headpool and sea-level were approximately equal at commencement, and level with the bottom of the channel. Sea-level was dropped during the run to encourage persistence of dynamic effects.

Controls: Medium sand; discharge 21.46 $cm^3 s^{-1}$; duration 35 minutes; headpool constant at 10.0 cm; sea-level at start 9.9 cm and at switch-off 6.1 cm, giving a fall of 3.8 cm or 0.108 cm/min.

Observations and data: Flow was turbulent along the entire channel. Immediately upon commencement of flow, erosion started in three distinct parts of the system: at the mouth, where water incised down to sea-level; in the channel bed, where the initial U-shape became modified to a rectangular section; and in the bends.

Water flowing round the bends undercut the concave bank, widening the channel, and bed ripples were produced. Sediment dislodged from the bank and the bed was deposited on the convex bank as an incipient point bar. These bar features were pronounced by MINUTE 10 (Figure 8.35).

The mouth of the stream was eroding strongly in the vertical plane, maintaining a graded profile, and also moved laterally, spreading deltaic sediments to left and right in a flagellating motion.

By MINUTE 20 the limbs B–C and D–E between the bends had become slightly asymmetrical (Figure 8.36).

Figure 8.34. Test bed for Experiment 5.

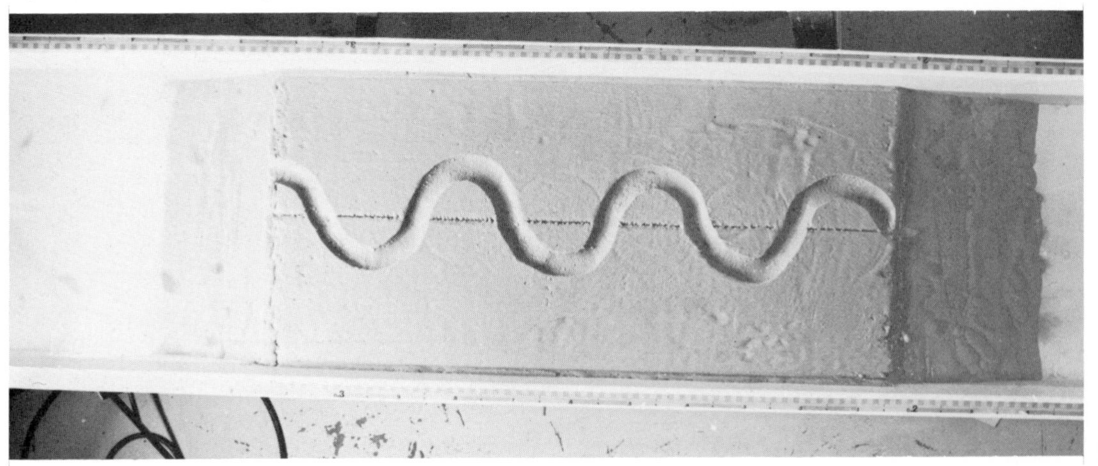

Figure 8.35. Experiment 5 at minute 10, showing incipient bar formation.

Figure 8.36. Experiment 5 at minute 20, showing slight asymmetry of limbs.

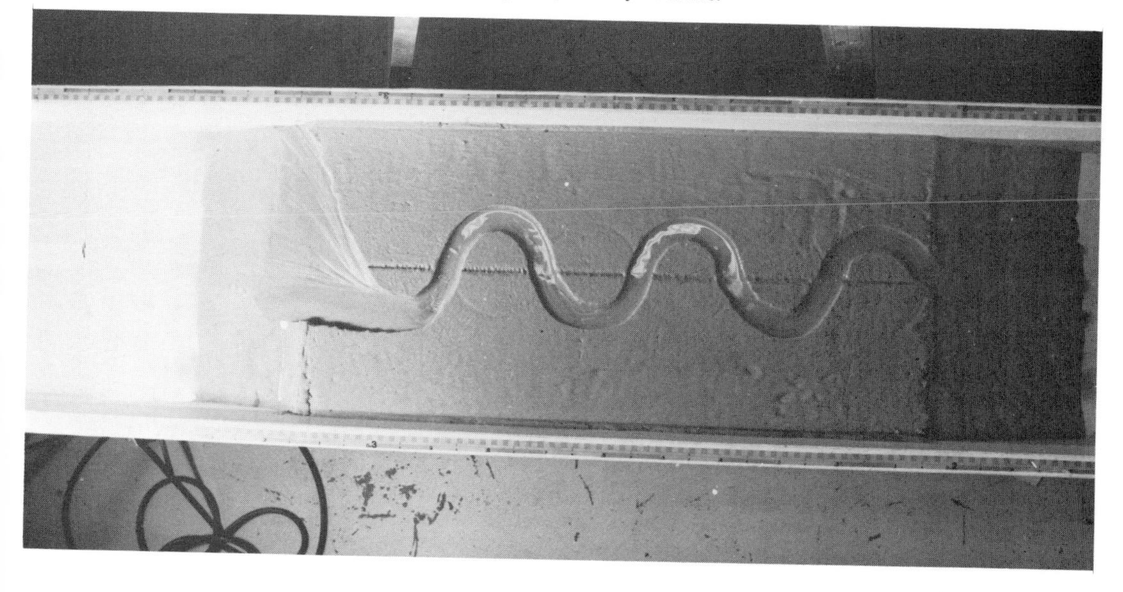

By MINUTE 25, sea-level was at 7.2 cm and the stream was cutting strongly into bend A, shifting the stream-line away from its original position (i.e. the bend was migrating), and deflected stream flow was carrying considerable debris to the delta (Figure 8.37).

At MINUTE 35, just at switch-off, bends B and C had slip-off slopes with rudimentary bars; limbs A–B, B–C, C–D, D–E and E–F were distorted by erosion with shift of stream-line, and bend A had an extensive slip-off slope on the convex bank. In general the concave banks of bends were steeper, undercut features, while the convex banks were gentler, slip-off features (Figure 8.38).

There was a variety of channel widths at switch-off, from 4.3 cm on bend D to 18 cm on bend A, which had widened to right and left of the test bed centre-line. Other bends were opened out about 5.0 cm from the original 4.2 cm, and the mouth was 37.5 cm across (Figure 8.39). Deltaic sediment weighed 830 g when dry.

Conclusion: The erosion on the bends in a stream is governed by the dynamics in other parts of the stream, and especially by the status of base-level.

Figure 8.37. Experiment 5 at minute 25, showing deflection of stream flow.

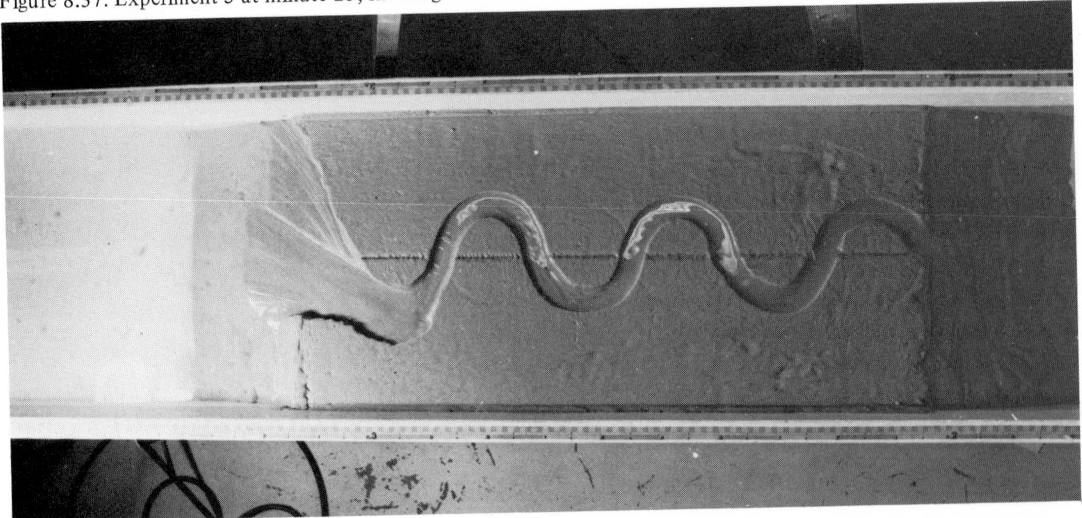

Figure 8.38. Experiment 5 at minute 35 (switch-off), showing final extent of erosion and deposition.

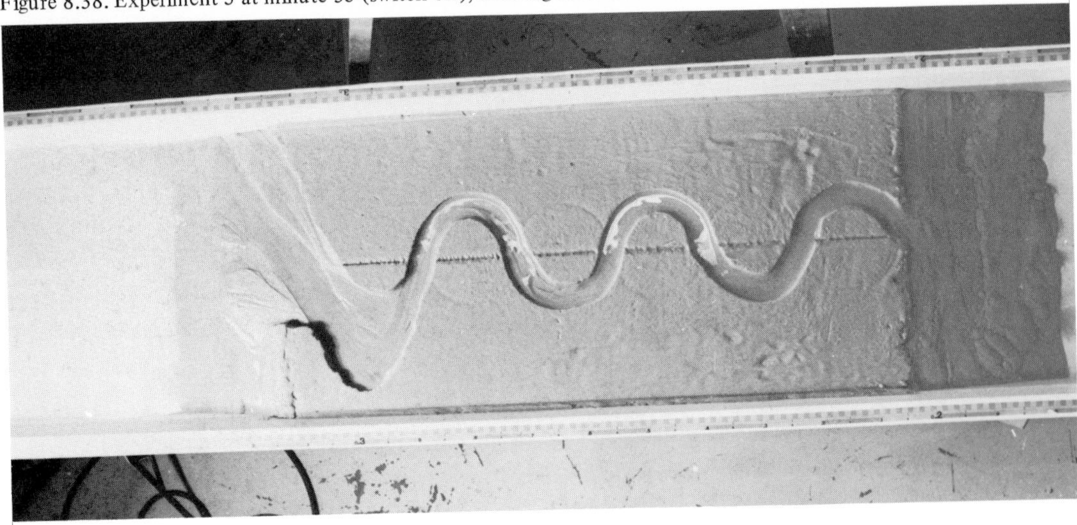

When the stream is approaching a graded condition, lateral erosion is strongest and bends, for example bend A, are rapidly cut into, producing a migration of the curve on at least the outer bank, but in the case of bend A, the convex bank also.

Further upstream, where lowering the base-level was probably only effective in slightly accelerating the water flow, lateral erosion on the bends was considerably less. The outer banks became undercut and material was transported to points where water swung across the channel, leaving bar deposits. These upper bends were in a state of temporary, or metastable, equilibrium, since one by one they would eventually be influenced by the steepening of the gradient as the nick-point travelled upstream deductively in a manner resembling that of bend A.

Systems interpretation: By design, the morphological subsystem at the commencement was too small for the cascading subsystem associated with it. This ensured that there would be some rapid modification of the morphological subsystem. The lowering of sea-level prevented the development of equilibrium conditions, certainly in the downstream part and apparently in upstream parts also. The process-response system was therefore one in which the channel form was continuously adapting to a constant and slightly too great mass input, and an ever-increasing energy throughput.

Figure 8.39. Schematic diagram of final channel features.

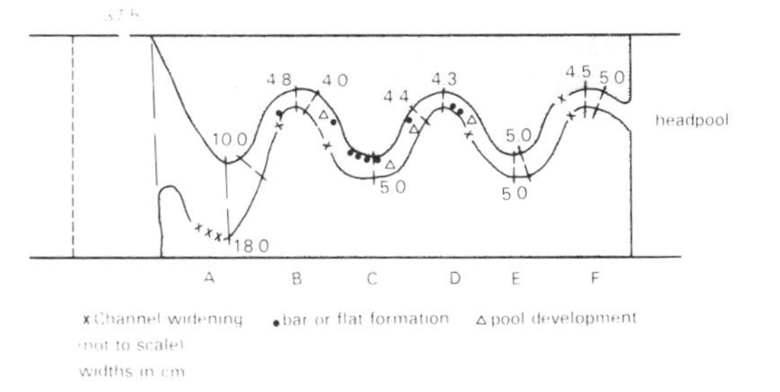

Modification of the morphological subsystem was coincident in three component parts: the mouth, the bed and the bends. The bed, after initial rapid adjustment, assumed a slow rate of change which included riffle, bar and incipient pool formation. The mouth, which first sensed the increasing energy throughput, assumed a rapid rate of change which, through bend A, influenced segments progressively further upstream.

The bends were, like the bed, mainly responding to the excessive mass throughput and associated high-energy throughput, with the great increment due to lowering sea-level mainly affecting bend A. This bend may be taken as an example of accelerated bend erosion and demonstrates that the final stage of bend migration is bend destruction.

The detailed process of erosion in a bend, by undercutting and transfer of sediment, represents a subsystem in its own right. In this case the water mass input is known, but the slope and energy throughput could only be approximately assessed.

The total channel system with 6 bends has a mass input of 45 066 g of water and an energy throughput of $45\,066(\frac{1}{2} \times 3.8 + 0.1)/10\,197.16$ J, where H is $\frac{1}{2}$(change of sea-level) + 0.1 cm, giving $E_p = 8.84$ J, together with 830 g of sediment with $E_p = [830 \times (\frac{1}{2} \times 3.8 + 0.1)]/(10\,197.16 \times 2) =$ 0.081 J, where $(\frac{1}{2} \times 3.8 + 0.1)$ gives mean H in the falling sea-level situation, and sand is dislodged from all positions along the channel, hence $\bar{H}$ for sand is $\frac{1}{2}(\frac{1}{2} \times 3.8 + 0.1)$.

The output from the system is 45 066 g of water, 830 g of sand, somewhat less than 8.921 J E_k, and heat energy (Figure 8.40).

If the subsystem considered is a single bend then mass throughput will be 45 066 g water + sediment from up-channel + increment by additional sediment from erosion in the bend. Energy can be based roughly on $\frac{1}{6}(\frac{1}{2} \times 3.8 + 0.1)$. If a single

Figure 8.40. Canonical diagram for channel system (Experiment 5).

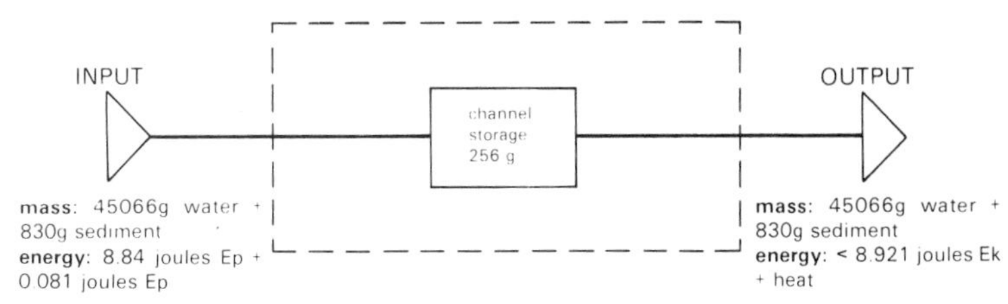

bend is required as a demonstration, the experiment should be re-designed to allow for easy measurement.

Experiment 6

Purpose: To determine the effect of the addition of sediment to a regular trapezoidal channel.
Apparatus: Flume; medium sand; sand hopper (Figure 8.41); former to emplace wide initial channel; timer, glassware etc.
Procedure: Medium sand was formed into a test bed measuring 1.17 m in length and containing a groove 25 cm wide and 2.4 cm deep (Fig. 8.42). Sea-level and headpool level were kept constant throughout the run. Two phases were employed, the first without addition of sand to produce a condition of equilibrium, and the second with addition of sand to produce perturbation and modification of channel form.
Controls: (Scale 1:1) Discharge constant at 526 $cm^3 s^{-1}$; sea-level constant at 8.6 cm; headpool level constant at 8.9 cm; duration of phase 1 without addition of sand 22 minutes, and of phase 2 with sand 52.5 minutes; sand added in phase 2 at a rate of 17.65 g/min at a point 5.0 cm from the headpool intake.
Observations and data (phase 1): Flow was turbulent. Almost immediately after commencement a standing wave pattern formed in the channel bed and banks were slightly eroded (Figure 8.43). At this stage there was no current ripple development but one or two lines of differentiated flow occurred, near the intake and at the right-hand side of the mouth. By MINUTE 19, just before cessation of phase 1, the standing wave pattern had given way to current ripples in the downstream part of the channel. Upstream, remnants of the initial standing wave pattern were still evident, and the bed exhibited much the same configuration as at MINUTE 5 (Figure 8.44).
Observations and data (phase 2): At MINUTE 11 of the second phase, current ripples on the bed were pronounced and filled the downstream 60 cm of the channel leaving the remaining 57 cm ripple-free.

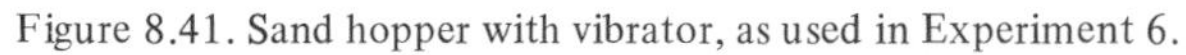

Figure 8.41. Sand hopper with vibrator, as used in Experiment 6.

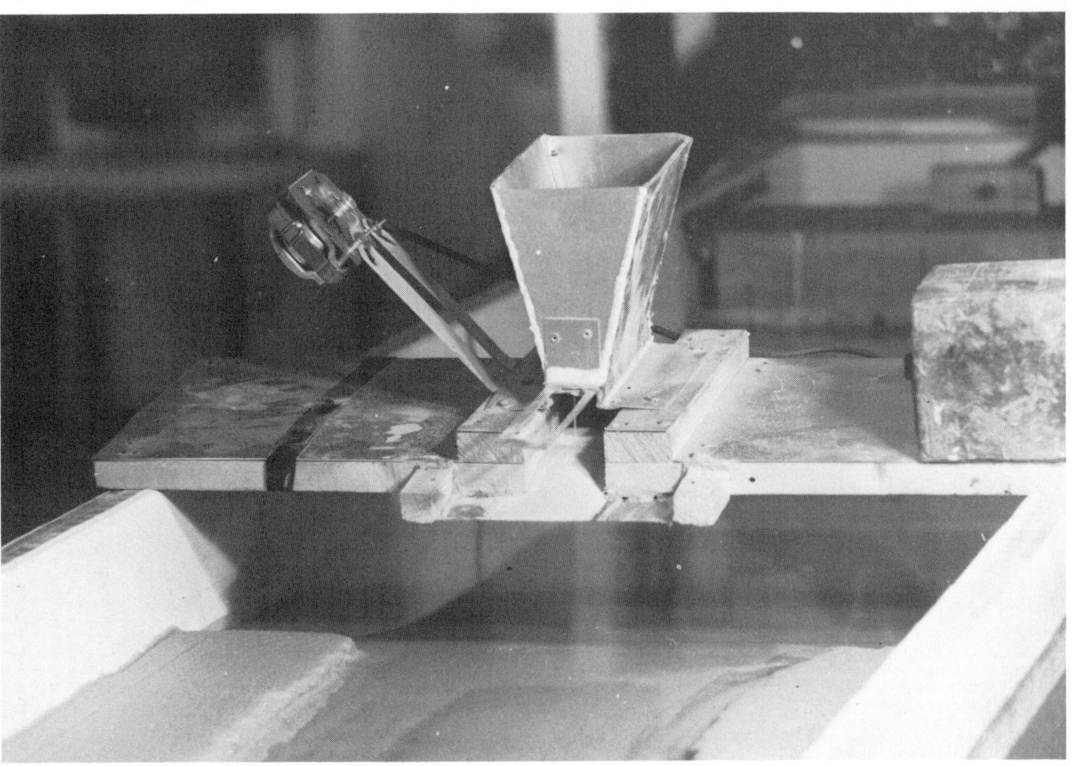

Figure 8.42. Test bed for Experiment 6.

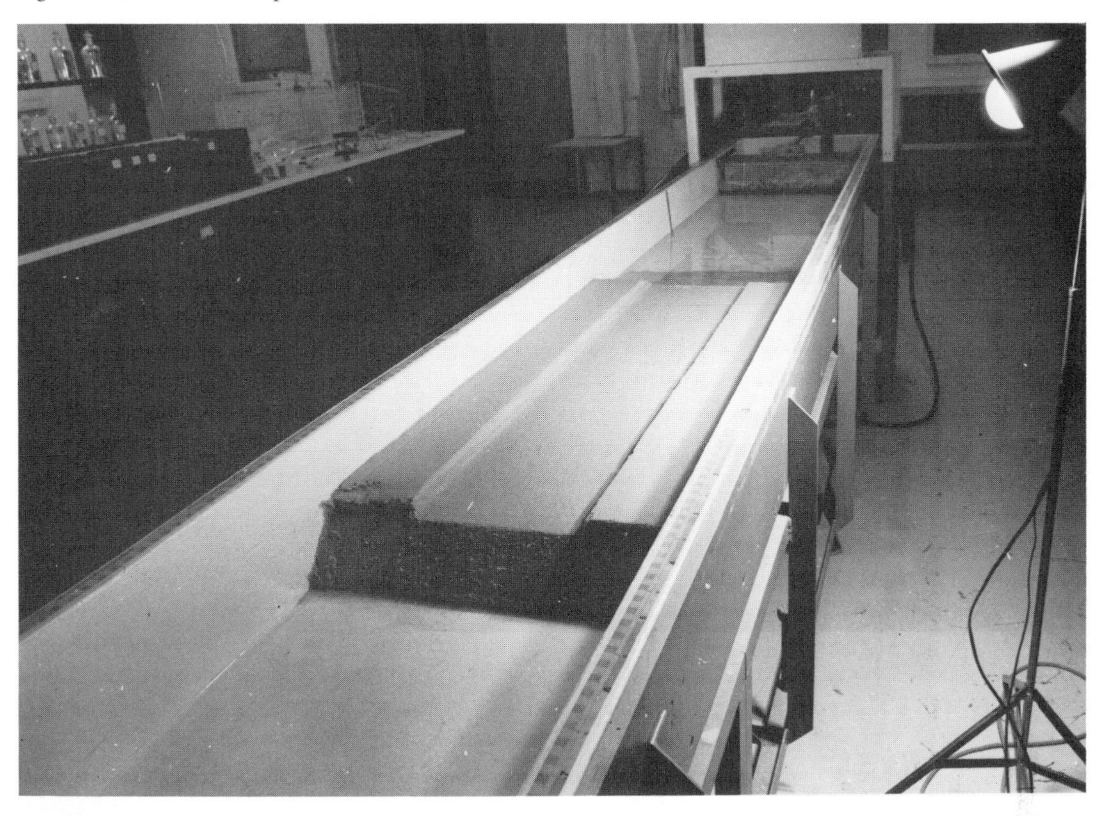

Figure 8.43. Experiment 6 soon after commencement, showing standing wave pattern.

Ripple spacing was fairly constant at 20 cm. At a point one or two centimetres downstream from the sand-feed a centre-bar had formed, leaving a section of smooth channel to either side where shooting flow occurred. The bar was oval in shape and, just downstream from its lower end, the left bank of the channel was eroded. Sediment moved downstream from ripple to hollow, to be deposited on the rapidly growing delta. Ripples were quite stable in form but moved slowly *en echelon* downstream (Figure 8.45).

Eventually the influence of the growing centre-bar dominated much of the channel. Bed ripples advanced downstream. Shooting flow around the bar or downstream from it, where the parted flow again united, was characteristic of most of the channel. Considerable lateral erosion occurred by flaring at the mouth and especially along the banks adjacent to the growing bar.

Most of the added sediment lodged in the bar or was deposited on the massive delta (Figure

Figure 8.44. Experiment 6 at minute 19 of phase 1.

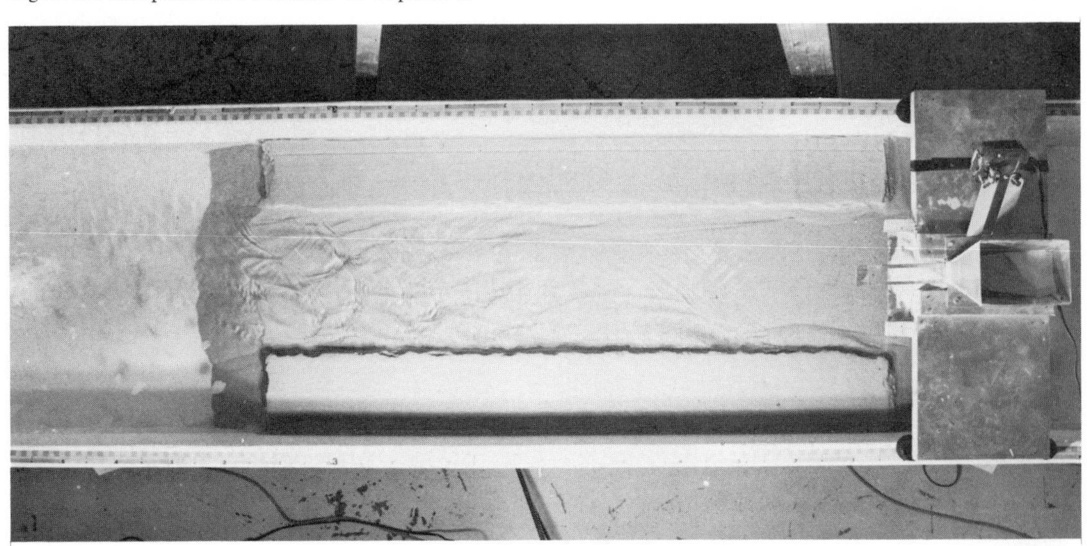

Figure 8.45. Experiment 6 at minute 11 of phase 2, showing ripple formation.

8.46). The final dry weight of sediment on the delta was 3127 g.

The process of bar formation appeared to be that sediment from the hopper landing on the bar remained on it until the sand almost reached the toe of the bar, then it moved off into lateral channels. Water, concentrated into the two lateral channels, eroded the banks and bank material was also carried downstream to aggrade the stream bed at the toe of the bar (Figure 8.47).

A post-run survey of the bed configuration showed the water depth over the bar to be 0.5 mm, while in the lateral channels to either side of the bar it was 1.1-2.2 cm (right channel) and 0.6-1.2 cm (left channel). Downstream in the centre channel depths were 1.9-2.0 cm, and at the delta the river depth was 0.5 cm.

In two places the stream channel had been gouged to depths exceeding the level of the channel mouth: at 25 cm upstream a depth of 1.0 cm below the channel mouth (1.5 cm below sea-level) was produced, and at 65 cm upstream a depth of 0.6 cm below the channel mouth (1.1 cm below sea-level) was produced (Figure 8.48).

Conclusion: The channel, which was pre-cut to approximately accommodate the discharge, was

Figure 8.46. Experiment 6 at end of phase 2, showing delta.

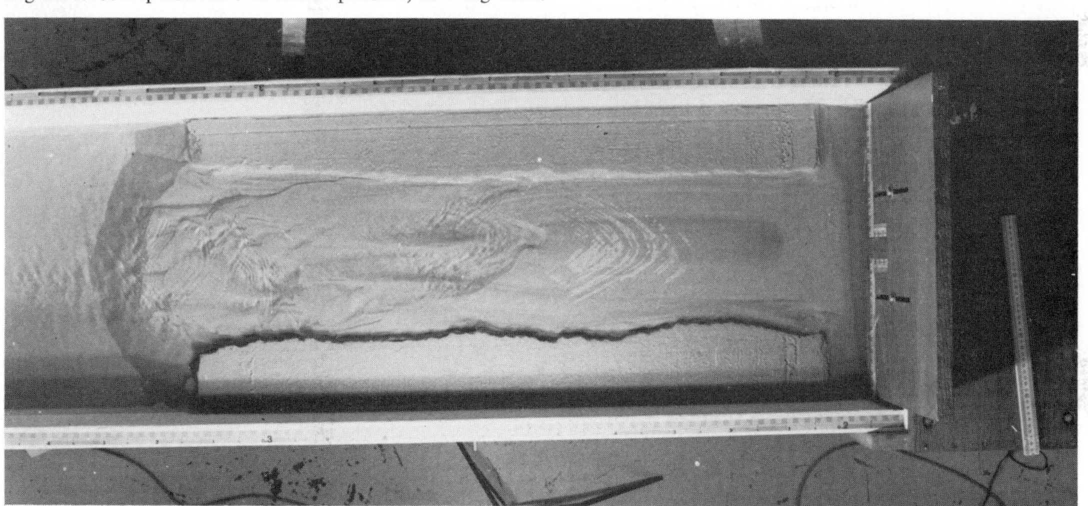

Figure 8.47. Particle trajectories from the banks at *a* and the bar at *b* and deposition on the stream bed to form a step.

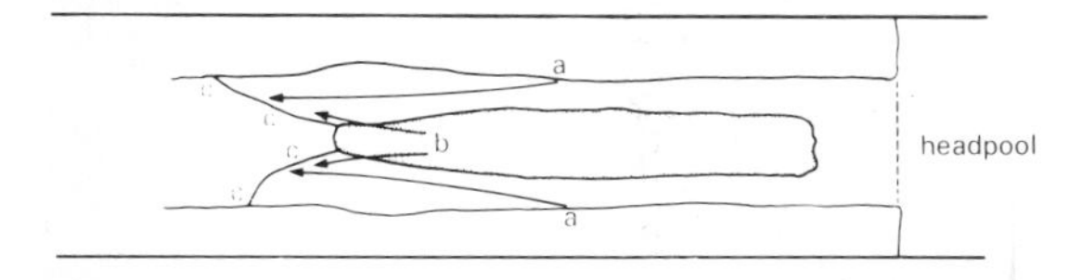

Figure 8.48. Location of pools in final stage of Experiment 6.

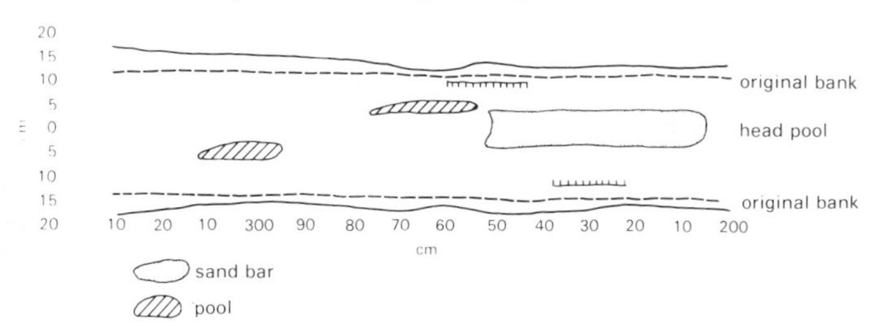

little modified in section as flow commenced. Banks were eroded minimally and there was little deltaic sediment. A characteristic of channel formation at this phase was current ripples which spread up-channel from the mouth. After commencement of sediment addition a bar began to form in centre-stream and stream flow parted into two segments, each of which developed into shooting flow and eroded the bed to either side of the bar. Lateral erosion of the banks occurred just downstream from the bar, especially on the left bank, and the bar slowly extended downstream, remaining slightly submerged. Particles swept off the bar were carried over the bed, to be added to the forward edge of an advancing bed-step which marked the forward limit of shooting flow. The current-rippled, streaming-flow section of the bed diminished in extent, and sediment moved over it from ripple to ripple until it eventually became deposited on the delta. A total of 927 g of sediment were added and 3127 g of sediment were deposited in the delta, therefore, since the elevation of the stream bed was constant, at least 2203 g of material were eroded from the banks, particularly from the mouth and in the proximity of the bar.

It is evident that bar formation affects flow mode, bed conformation downstream and to either side of the bar, and lateral erosion in the vicinity of the bar.

Systems interpretation: Assuming that the first phase, without sediment addition, was a process-response system in a near-equilibrium condition, the second phase may be regarded as producing a change in environment in which the initial system was perturbed. As is usual when the perturbing factor is continuous rather than spasmodic, the system became unstable and it did not approach a new equilibrium condition during the 52.5 minutes of duration of phase 2.

The inherited channel was approximately 25 cm in width and 0.9 cm mean depth (depth limits 0.5 and 1.1 cm), with current ripples over the downstream part of the bed and rudimentary delta formation.

In phase 2, the cascading subsystem was modified only by addition of 927 g of sediment at the rate of 17.65 g/min. The morphological subsystem became modified, firstly by the midstream accumulation of sediment and the subdivision of stream flow, and secondly by widening of the stream channel and changes in bed configuration.

The input to this perturbed system was: mass of water 526 g s^{-1} for 52.5 min = 1 656 900 g; mass of sand 926.6 g; energy [(1 656 900 × 0.3) + (926.6 × 0.3)]/10 197.16 = 48.77 J, together with sediment incorporated into the system by erosion.

As the quantity of sediment trapped in the bar is unknown, and the delta contained a small but unknown quantity of material from phase 1, neither input nor output for phase 2 can be precisely calculated.

The input and output for the total experiment are as follows.

Input – water mass 2 351 220 g, water energy 2 351 220 × 0.3/10197.16 = 69.17 J, sand E_p, based on deltaic material of 3127 g and mean channel-drop distance of 0.15 cm, 3127 × 0.15/10 197.16 = 0.046 J.

Output – mass 2 351 220 g of water + 3127 g of sand, with less than 69.216 J E_k + heat energy (Figure 8.49).

The morphological parameters at switch-off had changed to mean channel width of about 31 cm, mean depth of about 1.0 cm, but with greater diversity of deeps and shallows than in phase 1, together with the bar and associated parted channel-ways.

Figure 8.49. Canonical diagram for process-response system (Experiment 6).

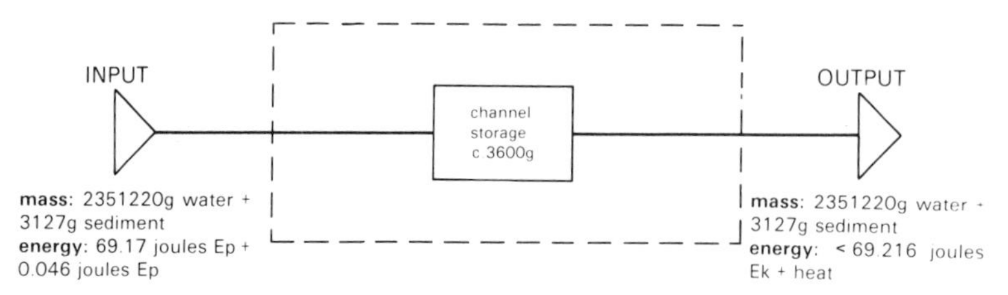

Experiment 7

Purpose: To produce meander cut-offs.

Apparatus: Flume with constant feed-system and falling sea-level control; carriage with plane former; wedges for slope forming; medium sand; circular former of 7.25 cm radius for curves; timer, glassware.

Procedure: A test bed of 1.06 m in length was formed with a 1 in 38 slope by pinning wedges to either wall of the flume (Figure 8.50). A central guide-line was scribed in the finished sand surface and three curves were marked out by scribing around the circular former. The channel was then carefully dug and trimmed to a width of 4.0 cm and a depth of 1.7 cm, and the sea and headpool were filled (Figures 8.51 and 8.52). The water input and sea-level drop mechanism were pre-set.

Controls: (Scale 1:1) Discharges 50.0 $cm^3 s^{-1}$; bed slope 1 in 38 (0.026% or 1° 30′); headpool level 12.9 cm; sea-level 10.0–9.2 cm, with a drop-rate

Figure 8.50. Wedges in position for production of sloping test bed for Experiment 7.

Figure 8.51. Test bed for Experiment 7.

of 0.063 cm/min; duration of experiment 12.67 minutes.

Observations and data: The stream began to form bars, pools and riffles almost from the commencement of flow, and to erode into the banks, especially at the outside of the curves, and by MINUTE 5 the channel had widened to 7.2 cm in limb A–B and curve C, producing strong asymmetry (Figure 8.53). By MINUTE 6, seepage across the neck of the land between bends A and C turned into a rill, through which flow coursed to form a gap and a cut-off at bend B (Figure 8.54). At MINUTE 8.5 water from the initial cut-off was beginning to flood over the neck of bend A (Figure 8.55), and by MINUTE 10 the second cut-off was fully formed (Figure 8.56). The stream course by this time was considerably shortened; by switch-off at MINUTE 12.5 the bend at A was carrying little or no water. while the bend at B was still an active channel. The conditions immediately after switch-off are seen in Figures 8.57 and 8.58, in which there has been considerable widening of the original 4.0 cm chan-

Figure 8.52. Experiment 7 immediately before commencement.

Figure 8.53. Experiment 7 at minute 5, showing commencement of asymmetry in curve C.

nel with consequent distortion of the symmetrical bend formation.

Conclusion: Stream flow energy was concentrated mainly into bank erosion, which, because of the near-graded condition of the channel from inception, was able to occur along the whole channel length. Erosion was particularly active on the outside of the bends and, where the bend was sufficiently convoluted to form a narrow neck, water seeped across to produce at first a rill and then a large channel. The resulting cut-offs were thus a consequence of a dual process, lateral bank erosion and seepage washout. By this series of events, the stream's main thalweg was shortened from 168.6 cm to 122.1 cm, which is a 27.6% reduction.

Systems interpretation: If sea-level had been constant, the stream system would have inherited a sloping morphological subsystem in a near-graded condition, and a channel cross-section a shade too small for the discharge. It would, on *a priori* evidence, have adapted its form, in some

Figure 8.54. Experiment 7 at minute 6, showing the breached neck in curve B.

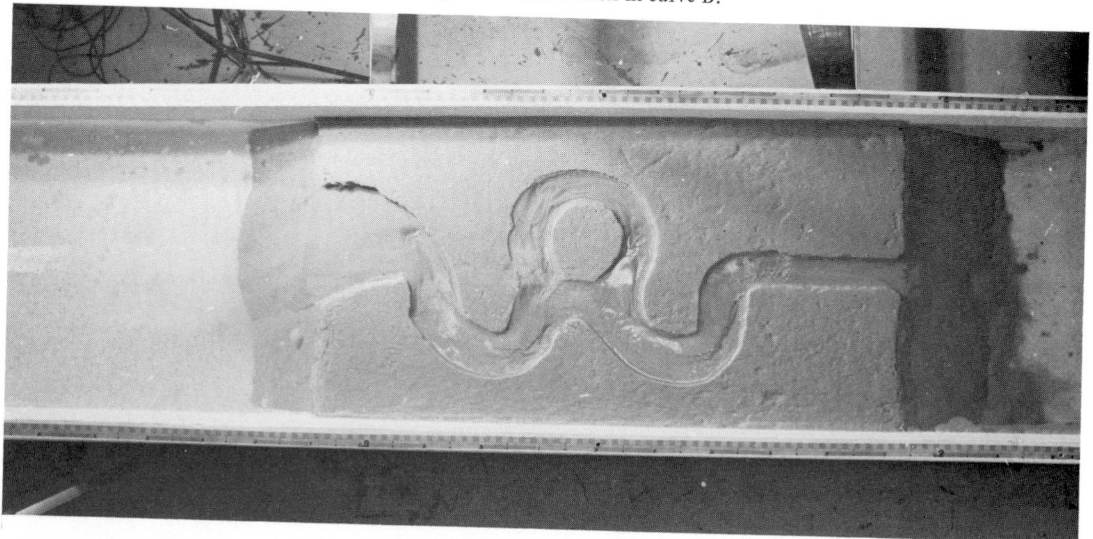

Figure 8.55. Experiment 7 at minute 8.5, with neck at curve A washing out.

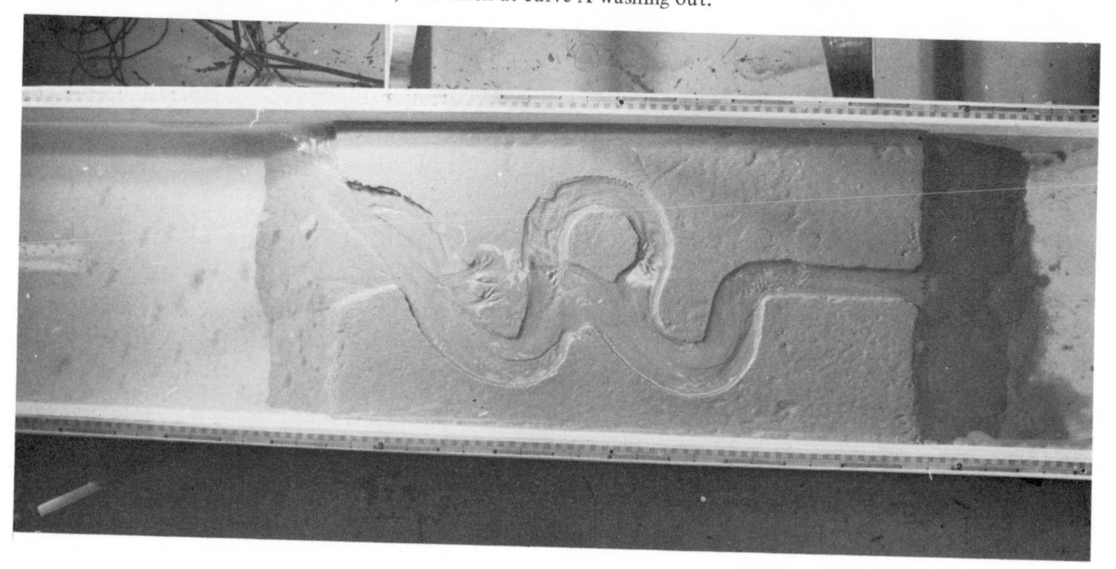

Figure 8.56. Experiment 7 at minute 10, with second cut-off fully formed.

Figure 8.57. Experiment 7 immediately after switch-off.

Figure 8.58. Schematic diagram of channel conditions after switch-off (Experiment 7).

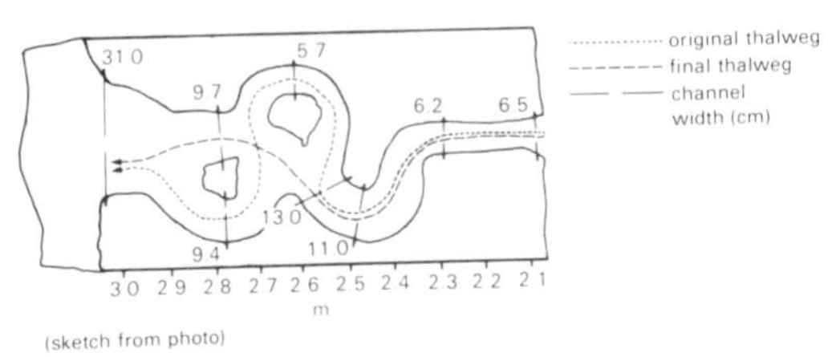

respects quite quickly and in others more slowly, to near-equilibrium. This would have represented recovery from an isolated perturbation by a negative-feedback loop. However, as sea-level was dropping throughout the experiment, inequilibrium was maintained and the constant increase in energy enforced constant adaptation of the morphological subsystem. Since a near-graded profile was part of the inherited form, energy was available throughout the experiment for lateral as well as vertical erosion.

The bends in the channel were inherited characteristics that were particularly vulnerable to lateral erosion, and by this process and seepage two of the bends were cut off. Since the bends were designed to accommodate only approximately the mass throughput, when cut-offs formed the resulting thalweg showed no tendency to curve in such a tight radius.

If the total channel segment is regarded as the process-response system, it may be seen as a segment comprising, initially, two straight limbs and three bends, and terminating as a straight link, a sharp bend, a sinuous downstream thalweg bordered by two cut-off channels, one of which still functioned to some extent as a routeway, and a coastal plain disgorging into a delta. The last morphological subsystem was not a final form since it was still rapidly changing. If the change were much slower, it might be proper to regard it as being in a condition of dynamic equilibrium, but retaining a 1:1 time scale it must be considered a condition of inequilibrium.*

* This is apparently true of the principal stages of all the laboratory models treated here, where the bed material is unconsolidated sand and the mass and energy factors used are calculated to produce readily perceived short-term effects.

Input to the stream system was water with mass $50 \times 760 = 38\,000$ g, and energy $38\,000 \times (2.9 + \frac{1}{2} \times 0.8)/10\,197.16 = 12.3$ J, plus sediment with mass 1507 g and potential energy $(E_p) = \frac{1}{2}(1507 \times 3.3)/10\,197.16 = 0.25$ J.

Output was 38 000 g of water and 1507 g sediment, with less than 12.55 J E_k plus heat (Figure 8.59). (Note that for energy input $H = 2.9$ cm = difference in elevation between headpool and initial sea-level $+ \frac{1}{2}$ sea-level drop distance $(\frac{1}{2} \times 0.8)$.) The storage capacity of the system, taking into account channel widening and the storage capacity of bends, increased from about 200 g at commencement to about 800 g at switch-off.

Experiment 8

Purpose: To produce a waterfall.

Apparatus: Flume with adjustable feed-system and controllable sea-level; medium sand; former; polythene sheet; timer; glassware.

Procedure: A test bed 1.35 m long was set up in two layers, the first was levelled off and trimmed and a piece of heavy polythene sheet 40 cm long and 32 cm wide was placed upon it with its forward edge 2.0 cm from the *cliff* of the model. The second layer of sand was then placed upon this, shocked firm and trimmed to 4.0 cm thickness, leaving the polythene sheet at 4.0 cm depth. A groove was emplaced and the whole surface was trimmed and brushed (Figure 8.60). Water feed and sea-level drop mechanism were pre-set to required values and sea and headpool were carefully filled. The sea was just within the guide channel at commencement (Figure 8.61).

Controls: Discharge 19.5 $cm^3 s^{-1}$; headpool level 10.0 cm; sea-level 10.0 cm at commencement, 2.25 cm at switch-off giving a drop distance of

Figure 8.59. Canonical diagram of cut-off system with morphological subsystem changing from highly curved to sinuous (Experiment 7).

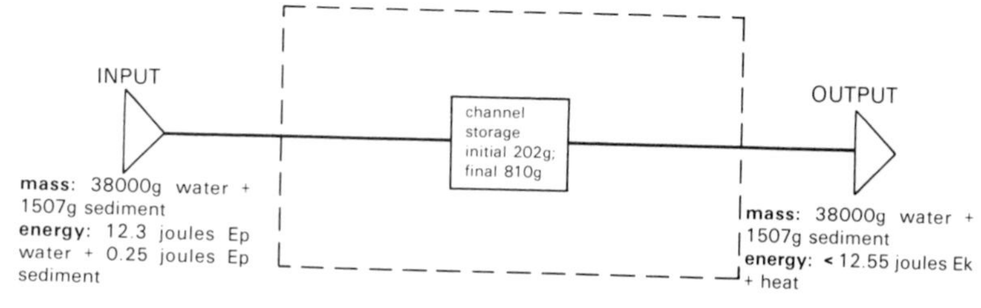

Figure 8.60. Test bed for Experiment 8.

7.75 cm with a drop rate of approximately 0.116 cm/min; duration of experiment 67 minutes. Scale can be 1:1. If the model is related to a chosen prototype, similitude will have to be considered.

Observations and data: For the first 30 minutes, erosion of the channel followed the normal process of incision as sea-level dropped, with a little bank undercutting and a small quantity of deltaic sediment being carried out (Figure 8.62). At MINUTE 33, when sea-level had fallen some 3.8 cm, the first suggestion of a resistant stratum began to appear in the stream bed, marked by a small ripple across the channel (Figure 8.63). Almost immediately afterwards a plunge-pool developed and a deep cavern was formed on the right bank at the fall (Figure 8.64). By MINUTE 46 lateral erosion had produced extensive bank collapse. Water leaving the plunge-pool parted to produce an island on the delta (Figure 8.65). By MINUTE 57 the waterfall was flowing in a concentrated stream from a centre part of the resistant

Figure 8.61. Experiment 8 at commencement, with sea just within the guide channel.

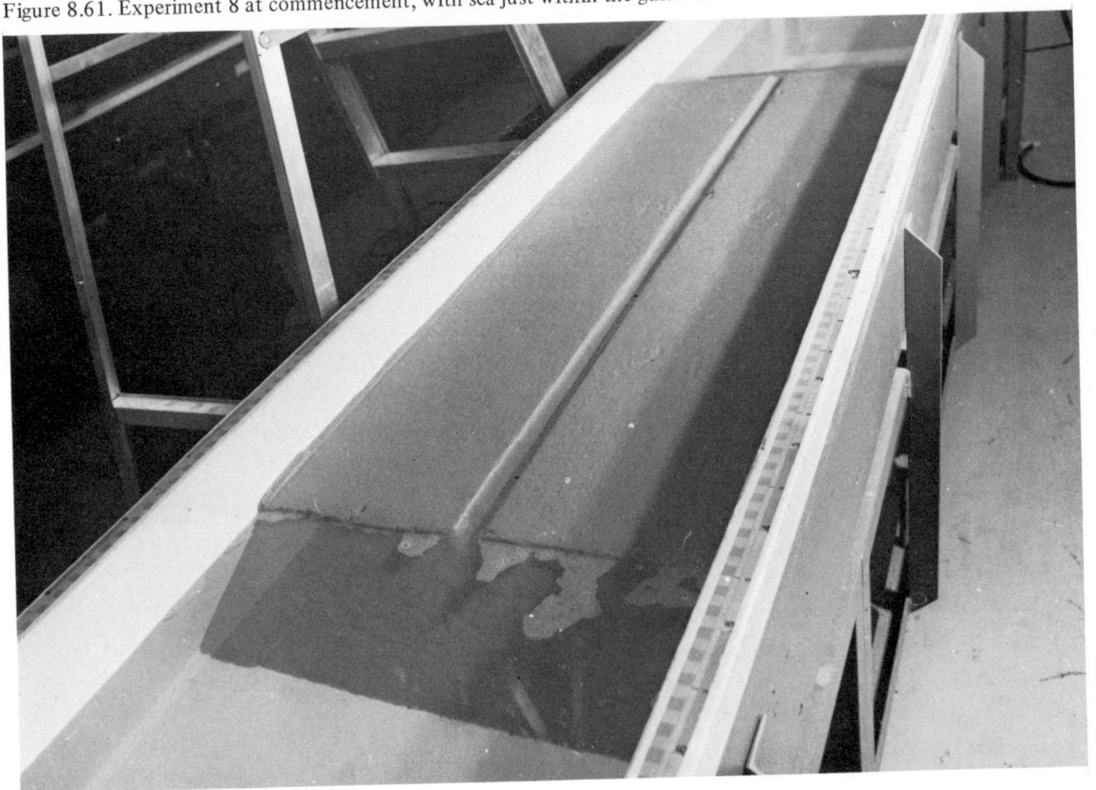

Figure 8.62. Experiment 8 after 30 minutes, with some incision and lateral erosion.

Figure 8.63. Experiment 8 at minute 33, showing first indication of a resistant bed layer.

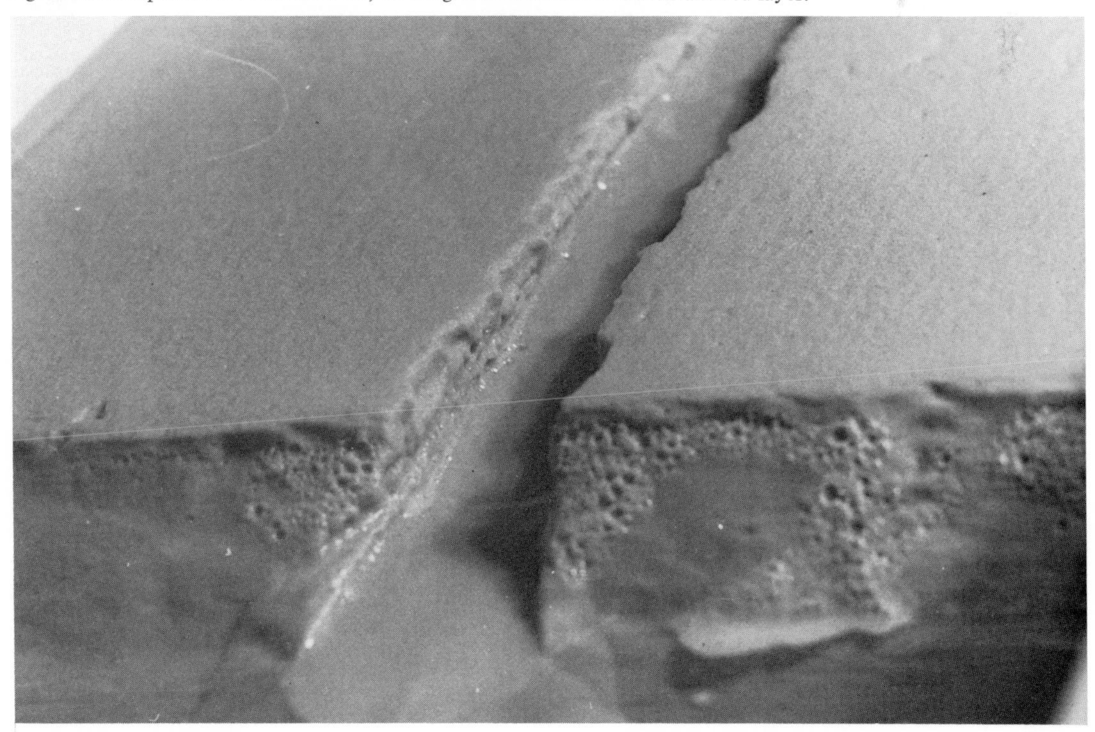

Figure 8.64. Experiment 8 almost immediately after minute 33, with plunge-pool and bank undercutting.

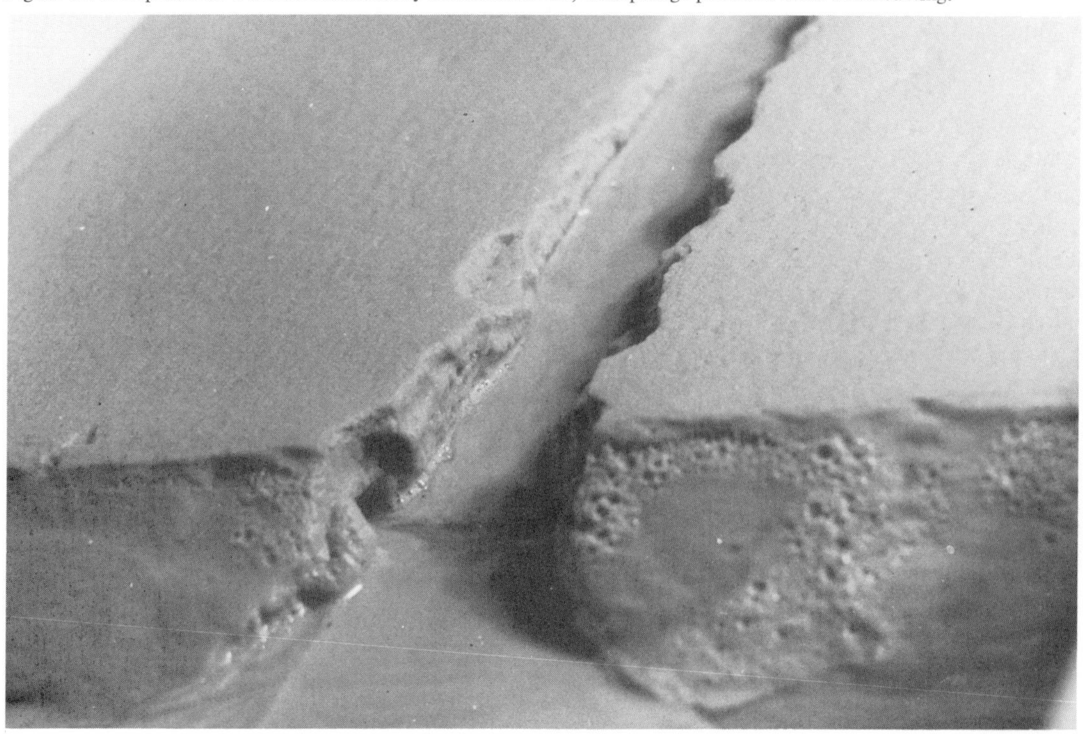

Figure 8.65. Experiment 8 at minute 46: deltaic island commencing.

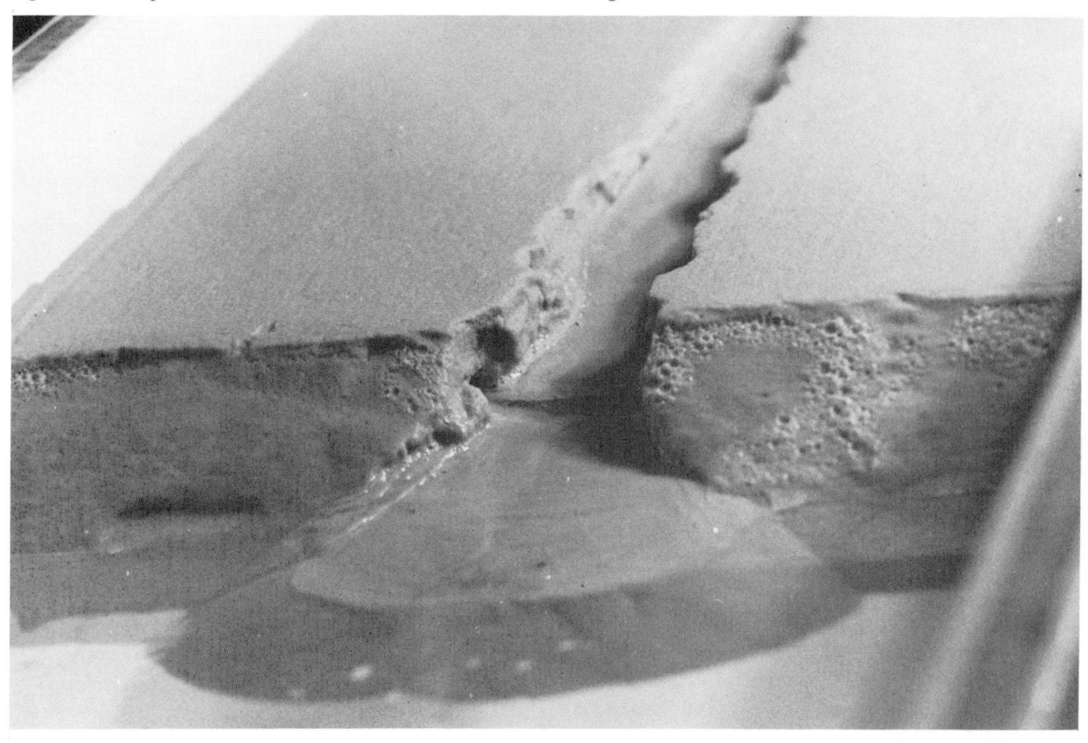

stratum. The plunge-pool was extensive and the right bank cavern greatly enlarged. By MINUTE 60 (Figures 8.66 and 8.67) the bank around the right bank cavern had collapsed and the delta island was extensive, and by MINUTE 67 at switch-off a very deep cavern existed beneath the overhanging resistant stratum and the deltaic island was a fairly high residual form (Figure 8.67). The final planimetric form is indicated in Figure 8.68, in which deltaic symmetry, the location of the fall, bank erosion upstream and the position of the nick-point are all clearly seen. The channel and valley remained coincident having a final width of 3.8 cm at 126 cm from the mouth. The canyon-form valley depth was 4.5 cm to the top of the fall and 6.5 cm to the bottom, and across the fall the stream/valley width was 13.5 cm. By switch-off the nick-point had moved 98 cm upstream and deltaic sediment weighed 2540 g.

Conclusion: The erosion of the channel was initially vertical as the stream cut down to a falling sea-level and lowering water table. This vertical incision lasted for 33 minutes, when the first sign of heterogeneity in the bed appeared as a ripple in the sand and a coincident bar of light on the water surface. The resistant layer was not apparent (Figure 8.63), so the channel flow must have been influenced by a swiftly-passing stage of the groundwater table easing below the impermeable-layer level just before the stream eroded the bed itself onto the resistant layer.

As sea-level dropped further, the base-level of erosion lowered with it as far as the edge of the resistant stratum, and upstream from that line the base-level was governed by the resistant layer. A waterfall developed within a few minutes of sea-level falling below the resistant rock, and a plunge-pool and backing cave also appeared.

Base-flow above the fall, aided by stream flow into the saturated basal deposits at fall-level, allowed extensive basal sapping to produce a deep cavern within 43 minutes of switch-on. This enlarged and capsized before MINUTE 60. Also, once the fall's resistant stratum began to function as an upstream base-level, lateral erosion increased and bank collapse occurred as far as the nick-point.

Figure 8.66. Experiment 8 at minute 60, with extensive collapse of right bank.

Figure 8.67. Experiment 8 at minute 67 (switch-off), showing deep cavern in right bank and high residual deltaic island.

Figure 8.68. Experiment 8 at minute 67 (switch-off). View of channel form and delta from above.

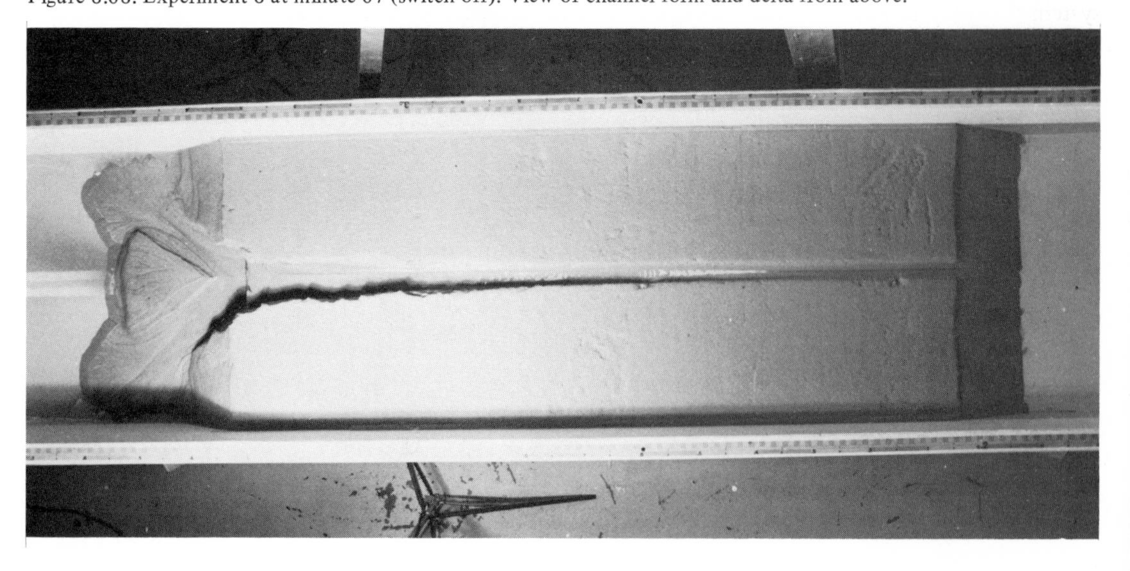

The plunge-pool was centred on the waterfall so that half of it lay beneath the lip of the fall, and an extensive cavern developed behind the fall (Figure 8.67). As water flowed from the pool it separated into two streams, forming a delta with an island, and, as sea-level lowered, the streams incised channels in the delta.

Systems interpretation: The development of a process-response system in homogeneous bed material with constantly falling base-level involves permanent inequilibrium. If sufficiently slow such a condition would be regarded as dynamic equilibrium, but this cannot be accepted in this situation, since analogues for theoretical analysis may not be used. Even so, the inequilibrium, caused by a steady increment of energy due to increase of relief, would be uniform and the first major change to be deduced would be when the nick-point eventually reached the headpool. After that it could be anticipated that the rate of lateral erosion would begin to increase upstream from the fall.

In the present experiment, after some 30 minutes running time, a differentiation of bed material was incorporated into the channel system, and a threshold was reached in which change in both morphology and cascade occurred. This was sudden enough to be regarded as a catastrophic event within the laboratory time scale and resulted in several features that could not otherwise have occurred. The new condition involved a stream system that possessed two distinct parts: one upstream from the fall, with a constant base-level of erosion, slowly retreating nick-point and strong lateral erosion; the other downstream from the fall, with a constantly lowering base-level of erosion, and dominant vertical erosion. The fall itself was in effect merely the boundary between these two subsystems. They were coincident in time and adjacent in space and may be represented canonically as in Figure 8.69.

The cascading subsystems may be represented from the emergence of the waterfall at MINUTE 30 if erosion prior to MINUTE 30 is estimated. The deltaic material was not removed at that time, in the middle of the run, but photo-analysis suggests a figure of about 10% of the final delta weight (10% of 2540 g) had been produced by MINUTE 30. The two subsequent systems therefore had cascades as follows.

(a) *System above fall*

Input: 37 minutes of flow at 19.5 $cm^3 s^{-1}$ = 43 290 g water mass. H is calculated using sea-level as it was at MINUTE 30 giving $H = 3.48$ cm, and E_p of water was therefore (43 290 x 3.48)/10 197.16 = 14.77 J. Sand mass is unknown but only a minor portion of the 2286 g of deltaic material is relative to the last 37 minutes (2540 − 10%). Sediment E_p is also unknown, but would be based on mass x $\bar{H}$ ($\bar{H} = \frac{1}{2} \times 3.48 = 1.74$ cm).

Output: Water mass 43 290 g; water energy (E_k) less than 14.77 J; sand mass x g; sand energy (E_k) less than 4 J; heat.

Figure 8.69. Stream process-response system for final 37 minutes of Experiment 8.

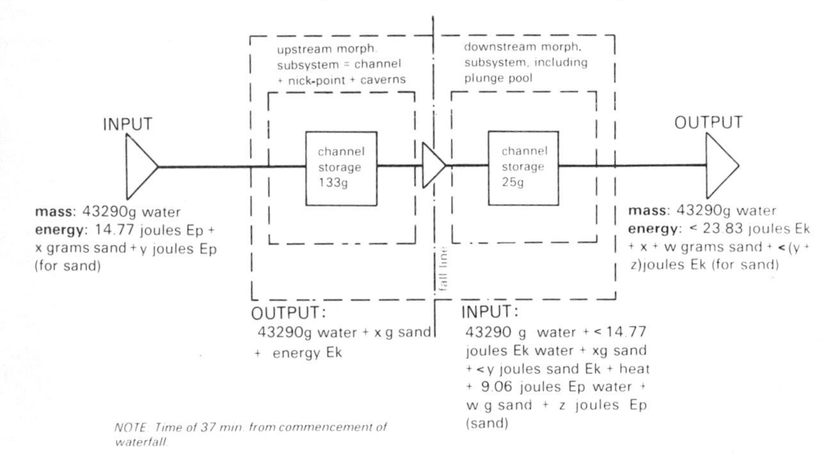

(b) *System below fall*
Input: mass = output from upstream system = 43 290 g water plus somewhat less than 14.77 J kinetic energy, due to loss in work done, plus heat energy, plus an additional quantity of potential energy due to the effect of water reaching the lip of the fall, and amounting to 43 290 g × H of fall (which is $\frac{1}{2}$ of 6.52 − 2.25 = 4.27 cm, where 6.52 cm is the sea-level at MINUTE 30 and 2.25 cm is the final sea-level, and $\frac{1}{2}$ this difference is taken since sea-level dropped steadily between these two values). E_p water input of lower system is therefore $\frac{1}{2}(43\,290 \times 4.27)/10\,197.16 = 9.06$ J. The mass input of sand is the material brought over the fall from the upper system plus material added by erosion within the lower system, totalling about 2286 g, the majority of which was derived from within the lower system. Potential energy of the sand input is also unknown but is a major part of the total sand mass potential energy, by reason of most sand originating below the falls and the $\bar{H}$-value of the lower system being greater than the $\bar{H}$-value of the upper system (2.135 cm, 1.74 cm respectively). Some heat exchange would also occur across the threshold between the upper and lower systems.
Output: from the lower system and, therefore, overall system: water mass 43 290 g, water energy 23.83 J. Sand mass 2286 g with kinetic energy of quantity unknown but of an order of magnitude of 1.0 J, which may be derived by taking 2286 g sand and an overall $\bar{H}$ for the two systems as 3.875 cm; $E_k \sim (2286 \times 3.875)/10\,197.16 = 0.87$ J. The kinetic energy output will be considerably reduced by plunge-pool erosion and hydraulic jump effect.

Experiment 9
Purpose: To produce stream beginnings at a spring-line.
Apparatus: Flume with controllable feed-system and sea-level; formers to give flat horizontal surface and a broad dihedral valley of flume width; flume side wedges; medium sand; gravel;* timer; glassware.

* In general, mixing of prepared and stored sand is avoided. If it is mixed it may have to be abandoned unless such grades are used that they can be easily separated by sieving.

Procedure: A bed of sand was made up and compacted and then levelled off with a straight former to give a horizontal plane surface. In this a valley was formed, sloping down-flume, with measurements 30 cm wide at the up-flume end, 55 cm wide at the down-flume end, 127 cm in length, with a slope of 1 in 41 to produce the upper surface of a dipping synclinal structure. At the up-flume end of this a 2.0 cm thick gravel layer was formed measuring 40 cm in length and taking up the synclinal shape (Figure 8.70). Upon this again was a further layer of fine compacted sand to give an escarpment of 5 cm maximum height above the valley head (Figure 8.71). The whole was carefully trimmed and brushed clean and the sea-level drop mechanism and water input were pre-set. Water was introduced at the headpool and the sea until the synclinal valley was flooded to the scarp foot on the shoulders (Figure 8.72). The scarp face was given the slope of its angle of rest underwater to avoid slumping into the sea at the valley head before the commencement of the experiment. The headpool was filled to the top of the escarpment and held constant by reference to a point gauge suspended from a metre-stick (best seen in Figure 8.72). This was necessary owing to the tendency

Figure 8.70. Test bed for Experiment 9 after formation of gravel layer.

of the headpool to lower if sea-level lowered. The run was started by unplugging the pre-set input pipe, and starting the sea-level mechanism and timer. Once the sea-level began to drop and water moved down the scarp face, the headpool level was monitored by the fine-adjustment inlet pipe at the constant-head tank.

Figure 8.71. Test bed for Experiment 9 after addition of final layer of sand.

Figure 8.72. Experiment 9 before commencement, with water introduced to scarp foot level.

Controls: Input varied from 0 $cm^3 s^{-1}$ at commencement to 11.6 $cm^3 s^{-1}$ at switch-off, with constant attention being paid to maintaining headpool level on the point gauge (12.7 cm); sea-level varied from 12.7 cm at commencement to 2.6 cm at switch-off, giving a drop of 10.1 cm at a rate of 0.124 cm/min; duration 81.5 minutes. Assume a scale of 1:1000.

Observations and data: As soon as the sea-level began to fall the synclinal structure began to emerge as a sea-filled bay, and the permeable interbedded stratum of gravel began to appear. At MINUTE 5 some minor slumping from the plateau face had occurred. By MINUTE 9 the sea had left the shoulders of the syncline and seepage from the gravel layer was apparent at 10 cm from both left and right walls, marking the first appearance of springs. This was accentuated by MINUTE 15 and at MINUTE 18 the flow of spring water was pronounced (Figure 8.73). By MINUTE 22 the water in the bay had fallen by 2.7 cm and spring-produced sediment was curving to follow the retreating shoreline. By MINUTE 25 (Figure 8.74) springs had produced stream channels and debris was deposited in plumes towards the axis of the bay. Evidence of spring seepage appeared towards the centre of the bayhead shoreline. By MINUTE 31 the bay-water had retreated sufficiently to expose much of the bayhead synclinal and the two initial streams began to coalesce. Four more seepage points emerged at the foot of the escarpment (Figure 8.75). By MINUTE 38, at 24 cm from the left wall, upwelling from a spring was strong enough to suggest artesian pressure. At MINUTE 42 (Figure 8.76) a centre stream began to emerge at the artesian spring, producing a new gravel exposure. By MINUTE 45 the new centre stream had become a preferred route for seepage and the original left lateral had almost ceased to flow. By MINUTE 52 the centre stream was dominant. The left lateral was almost defunct; the right lateral was flowing weakly. At MINUTE 78 an ink test showed some seepage from the left spring as well as the right. The centre spring was strong and flow in the stream was turbulent. The main stream had incised a clear channel downstream and some sinuosity had developed (Figure 8.77). By MINUTE 81.5 at switch-off (Figure 8.78) the spring-line had developed into a main spring to

Figure 8.73. Experiment 9 at minute 18, with spring-water flow increasing.

Figure 8.74. Experiment 9 at minute 25: springs have produced stream channels.

left of the centre-line, measuring about 11.0 cm in width, with two subsidiary springs each of about 4.5 cm in width. (The two outer gravel patches were on the shoulder of the syncline above the water table.) Seepage from the spring-line moved into the stream channel across a broad area of anastomosing filaments some 20 cm in width and 1.8 cm in depth. The downstream segment comprised a valley which, 5 cm upstream from the mouth, measured 9.5 cm in width and 1.1 cm in depth containing a stream of 7.5 cm in width and 1 mm in depth. Deltaic sediment had a dry mass of 490 g.

Conclusion: While the sea-level was coincident with the headpool level a water table developed within the highly permeable gravel layer. The active part of the experiment commenced as soon as sea-level began to drop. No appreciable seepage occurred until 9 minutes of the run had elapsed and sea-level had dropped by 1.1 cm, when it was necessary to maintain an input to the headpool of 4.0 $cm^3 s^{-1}$ which must, therefore, have been the

Figure 8.75. Experiment 9 at minute 31: seepage points have increased and streams are coalescing.

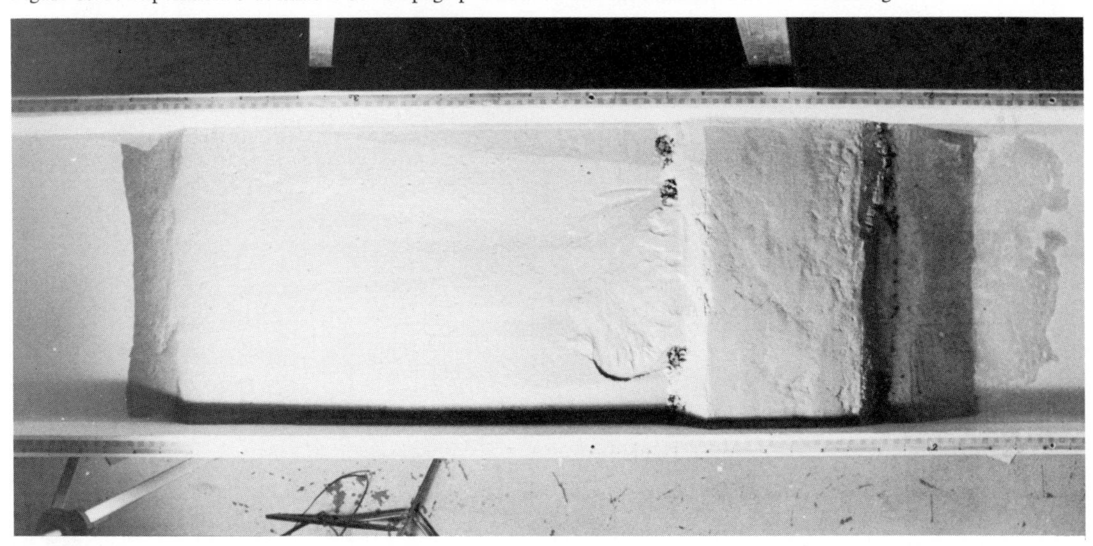

Figure 8.76. Experiment 9 at minute 42, with a central stream forming at the artesian spring.

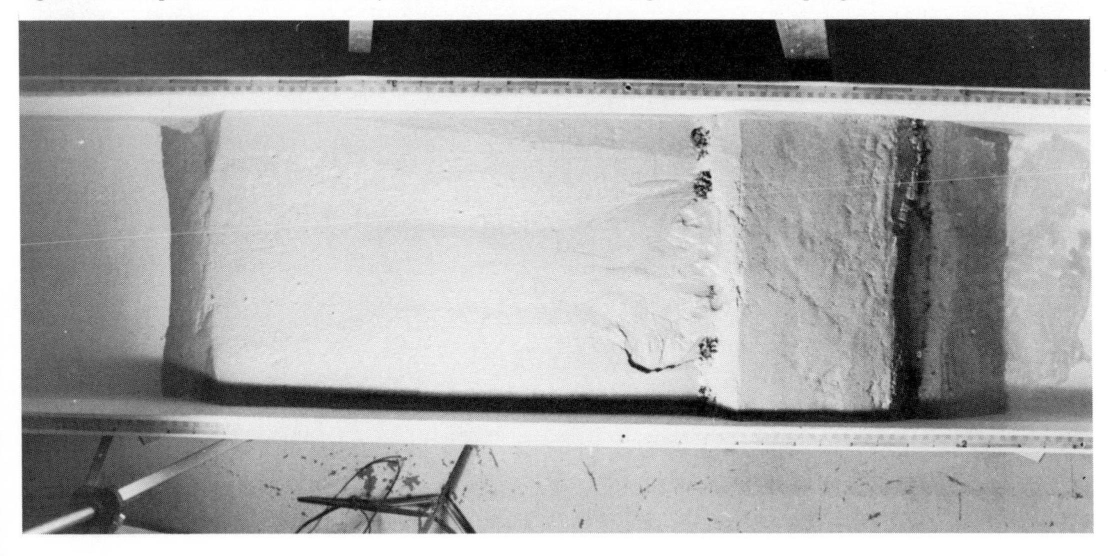

value of the basal flow through the gravel bed. After 15 minutes running time gravel was well exposed in left and right locations, demonstrating the existence of springs to either side of the syncline where a head of water existed between the headpool and those positions in the limbs of the gravel layer synclinal at sea-level and close to the receding bayhead. In these locations water would concentrate from higher up the limbs, flowing at the base of the gravel over the underlying sand surface. These two springs produced streams that ran along the bay shore for some distance depositing sediment laterally into the bay, and extending as the bay shore retreated (Figure 8.74). As sea-level dropped the hydraulic head increased considerably and, after 38 minutes of running time, upwelling at the centre of the scarp face demonstrated the position of maximum head and strong-

Figure 8.77. Experiment 9 at minute 78: centre spring is strong with some turbulence in the stream.

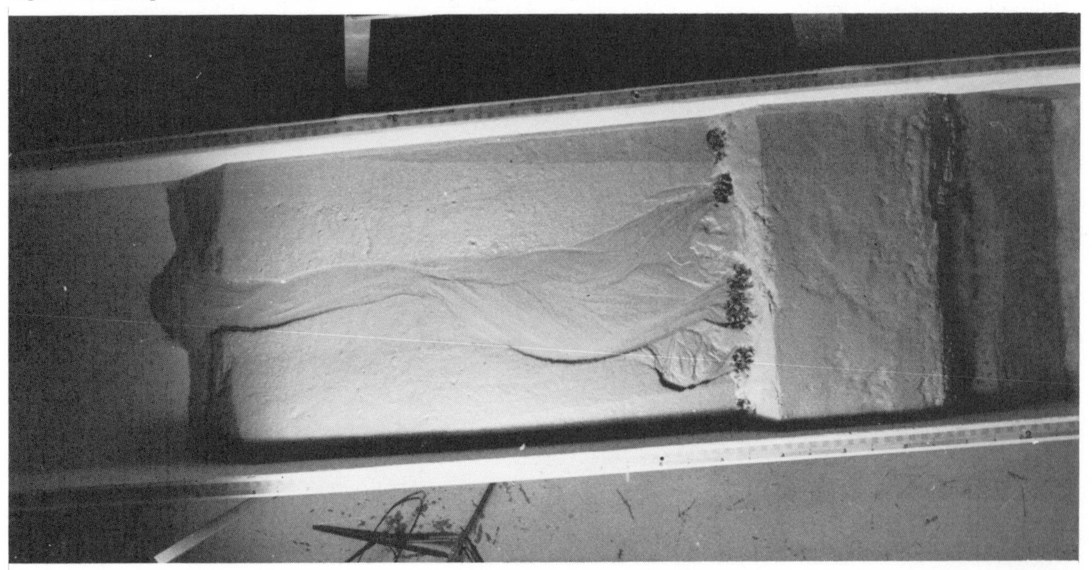

Figure 8.78. Experiment 9 at minute 81.5 (switch-off), with a main spring and well-exposed gravel layer. The sea has entirely retreated from the synclinal valley.

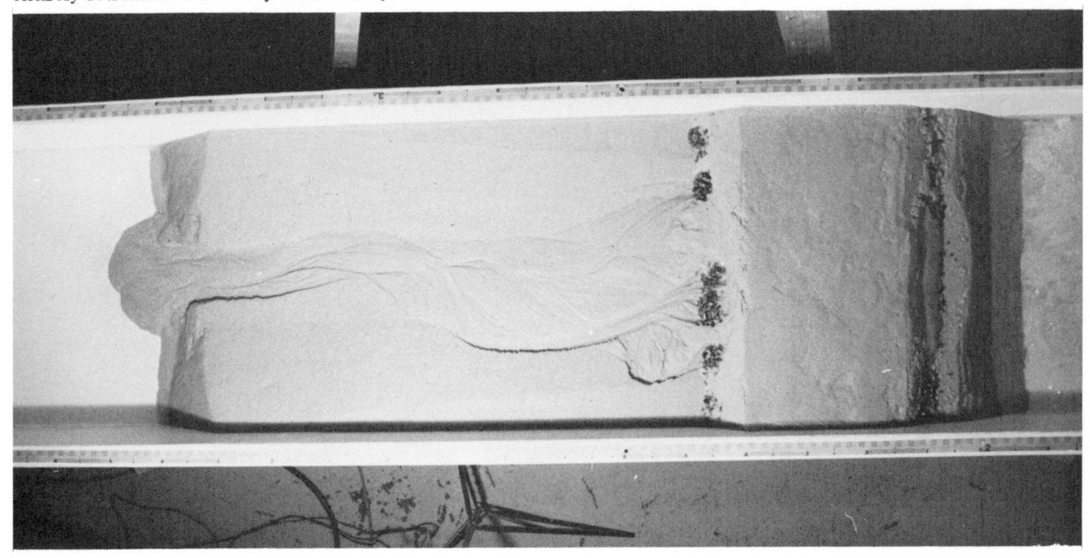

est seepage, and 15 minutes later a centre stream had formed and become dominant. The laterals had by this time begun to atrophy as water moving through the gravel bed sought a route to the lowest point. Slumping of the scarp face occurred over the spring zone and spring water washed in anastomosing filaments over the sand surface of the emerged bayhead, where the water table was coincident with the land surface. This, in a natural environment, would probably be a swamp zone. Down-valley, where the water table was eventually able to drop below ground level, a deeper inner valley occurred, containing a definite stream.

Systems interpretation: There are two principal systems in the experiment, the escarpment that backs the spring-line and the stream that drains the emerging syncline. As various aspects of stream systems have been considered in other experiments, only the escarpment will be considered here (Figure 8.79).

Before commencement, the gravel layer acted as a groundwater store of about 1700 cm^3 gross volume (porosity and permeability not determined). After commencement, mass input slowly increased to a maximum at switch-off of 11.6 $cm^3 s^{-1}$, to give a mass input of $\frac{1}{2} \times 11.6 \times 81.5 \times 60 = 28\,362$ g of water, and energy input was 28 362 g $\times \frac{1}{2}$(fall in sea-level against the scarp face)* (2/2 = 1 cm) or (28 362 × 1)/10 197.16 = 2.8 J. Very little modification of the morphological subsystem occurred, so that mass output at the interface between escarpment and stream systems (the spring-line) consisted of water without sand, with less than 2.8 J kinetic energy plus heat energy. The output of the escarpment system became the input of the stream system, representing a typical example of a *cascade*.

The escarpment system may be regarded as one in metastable equilibrium, since its stability depends upon the continuation of the system it

* It is not correct to use total sea-level drop because below scarp foot level the stream system utilizes the sea drop factor and the head of the stream system is the base of the escarpment system.

Figure 8.79. Escarpment process-response system (Experiment 9).

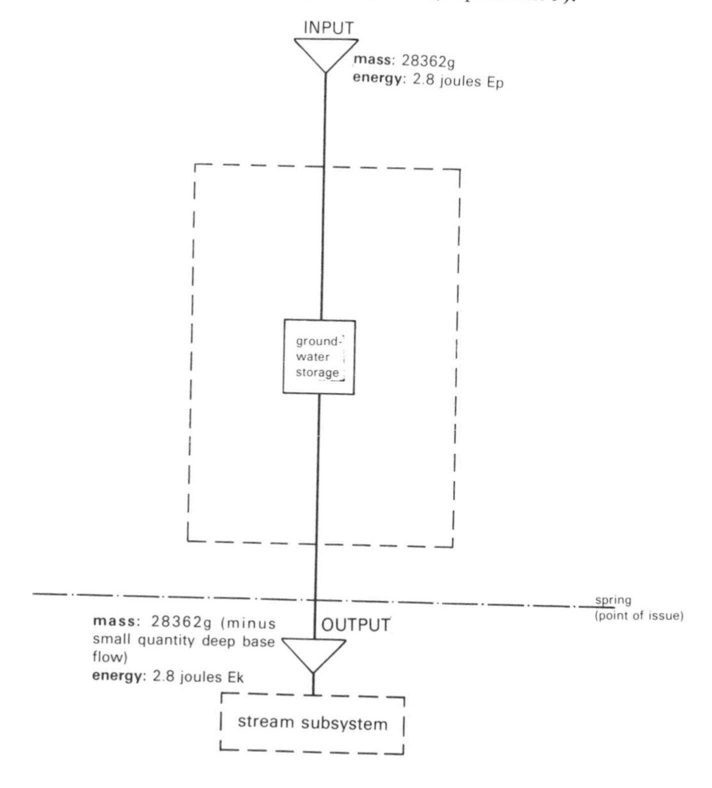

feeds. Once the stream system's modifying long profile changes at the scarp foot, the springs will change catastrophically to enter a phase of instability and, later, dynamic equilibrium.

8.3 Experiment 9 as a scaled model (discussion)

Previous experiments in this chapter have used a 1:1 scale, implying that processes and forms demonstrated are as applicable at the laboratory scale as they would be on a larger, out of doors scale. Strictly speaking, this same attitude may be taken with respect to the laboratory formation of a spring-line, except that terms such as scarp face and synclinal may be adjudged to belong to a larger prototype. In the present example the model is expanded to a feasible prototype, rather than the reverse as in most engineering models.

The scalar attributes are (*a*) geometric, (*b*) kinematic and (*c*) dynamic (refer to Chapter 3), each of which is now considered in turn.

8.3.1 Geometric similitude. Since the model does not represent a specific landform, any suitable scale may be adopted for linear parameters. Let vertical and horizontal scales both be 1:1000.

This will give a scarp height at commencement of 30 m (3 cm); a drop in sea-level of 101 m (10.1 cm); a final total height of escarpment above sea-level of 121 m (12.1 cm); a height of escarpment above foot of escarpment at head of synclinal of 50 m (5 cm); and at switch-off at MINUTE 81.5 (time scale to be decided upon) main spring zone 110 m wide (11.0 cm), subsidiary spring zones 45 m wide (4.5 cm).

In the synclinal portion of the model the area of anastomosing channels is 200 m wide (20 cm), occupying a valley of 200 m width (20 cm) and 18 m deep (1.8 cm), becoming, nearer to the sea, a stream 75 m wide (7.5 cm) and 1 m deep (0.1 cm), in a valley 95 m wide (9.5 cm) and 11 metres deep (1.1 cm).

Deltaic sediment has a dry volume of 318 cm^3 which, in the prototype, will be a volume of $3.18 \times 10^{11} cm^3 = 3.18 \times 10^5 m^3$. Water input volume: the model has an input of 28 362 cm^3 during the experimental time period, therefore the prototype will have an input of $2.8362 \times 10^{13} cm^3$ (or $2.8 \times 10^7 m^3$) during its life.

Bed material may be regarded as having a 1:1 scale, since sand and gravel sizes could as easily occur in the prototype as in the model. If the sizes of bed material particles are increased in the scale 1:1000 the prototype bed materials will be too large for credibility.

8.3.2 Kinematic similitude. The time span for the prototype will affect such attributes as the rate of sea-level drop and the rate of mass and energy input to the prototype system. A credible time interval for rate of sea-level drop (or land uplift, depending upon whether a eustatic or isostatic process is envisaged) would be 10 000 years for 101 m. This would be commensurate with post-glacial sea-level changes. However, with a total water input of $2.8 \times 10^7 m^3$ the system would receive water at the rate of 2800 m^3/yr or 7.7 m^3/day or $8.87 \times 10^{-5} m^3 s^{-1} = 88.7 cm^3 s^{-1}$, which would be too low for this kind of prototype landform – it would not produce a spring-line nor erode adequately to maintain the channel forms described above.

If the prototype time span is reduced to 100 years the rate of flow will be $2.8 \times 10^5 m^3$/yr or 767 m^3/day or 8877 $cm^3 s^{-1}$, which is still low in erosional terms and additionally requires a very high rate of sea-level drop (1.01 m/yr).

It is clear that this model, when converted to a prototype, requires a compromise time scale, and 81.5 minutes to 100 years (1:645 350) may be the best choice for this discussion.

Since the horizontal and vertical scales in the model are equal, trajectories of moving particles, for instance the trajectory of water particles moving through the gravel stratum or through stream channels, will be similar in model and prototype.

Velocities in the prototype would be proportionally less than those in the model. For instance, the velocity of flow through the permeable gravel layer is calculated as follows for the model: quantity of water in 81.5 minutes is 28 362 $cm^3 = 5.8 cm^3 s^{-1}$, cross-sectional area of saturated portion of gravel layer is 42.6 cm^2, therefore average velocity of flow is 5.8/42.6 =

0.136 cm s^{-1}; and for the prototype: quantity of water in 100 years is 2.8362×10^{13} cm^3 = 8987 $cm^3 s^{-1}$, cross sectional area of saturated portion of gravel stratum is 4.26×10^{10} cm^2, therefore average velocity of flow is 8987/ $(4.26 \times 10^{10}) = 2.1 \times 10^{-7}$ cm s^{-1}. Thus the velocity ratio is 1:647 619 (prototype:model). The ratio of average velocities of water in the synclinal streams may be calculated in a similar way.

8.3.3 Dynamic similitude. The primary factor to be considered is input of mass and energy. The mass input of water will be determined by its volume, and, if temperature is ignored, this will be $2.8 \times 10^7 m^3 = 2.8 \times 10^{13} cm^3 = 2.8 \times 10^{13}$ g.

Potential energy, which must take the gravitational factor into account, is a function of hydraulic head in the escarpment, which is $\frac{1}{2} \times 20$ m or $\frac{1}{2} \times 2000$ cm = 1000 cm = $(\bar{H})$.

Hence, energy is $m\bar{H}/10\,197.16$ = $(2.8 \times 10^{13} \times 1000)/10\,197.16 = 2.75 \times 10^{12}$ J.

Other factors such as sediment mass and energy (negligible), viscosity and capillarity, may be ignored as they are of a different order of magnitude.

Experiment 10

Purpose: To produce a river delta.

Apparatus: Flume; former to give a flat, horizontal surface with central guide-groove; medium sand; timer; glassware.

Procedure: The water feed-system was pre-set. A bed of sand 1.21 m in length and 6.0 cm thick was made up and compacted and then levelled off with a straight former and scribe to give a flat surface with a central V-groove. The bed was trimmed and brushed clean. The sea was introduced to within 1.0 cm from the bottom of the groove and the coastal profile was marked with grease pencil on the side windows. The headpool was filled to the bottom of the V-groove (Figure 8.80) and the run was started by unplugging the pre-set feed pipe in the constant-head reservoir. Velocity measurements were taken with dye slugs over 1.0 m segments.

Controls: Slope of channel during run 0.3 cm in 1.22 m (1:407); discharge constant at 35.4 $cm^3 s^{-1}$; sea-level constant at 5.0 cm depth; duration 43.3 minutes. Assume a scale of 1:1.

Observations and data: Within a few seconds of commencement the stream had produced a rectangular section in the lower reaches and material was carried to the mouth; by MINUTE 5.5 (Figure 8.81) a small delta had begun to form. By MINUTE 21 distribution lobes had developed on the seaward face of the delta, at an angle of rest parallel to the

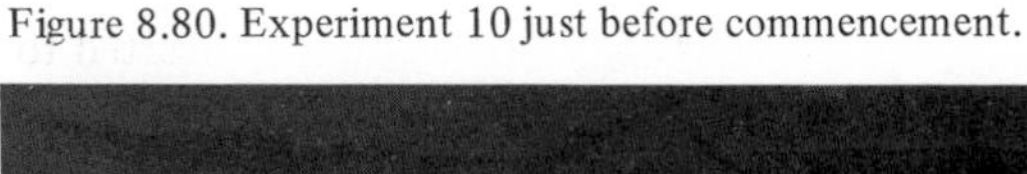

Figure 8.80. Experiment 10 just before commencement.

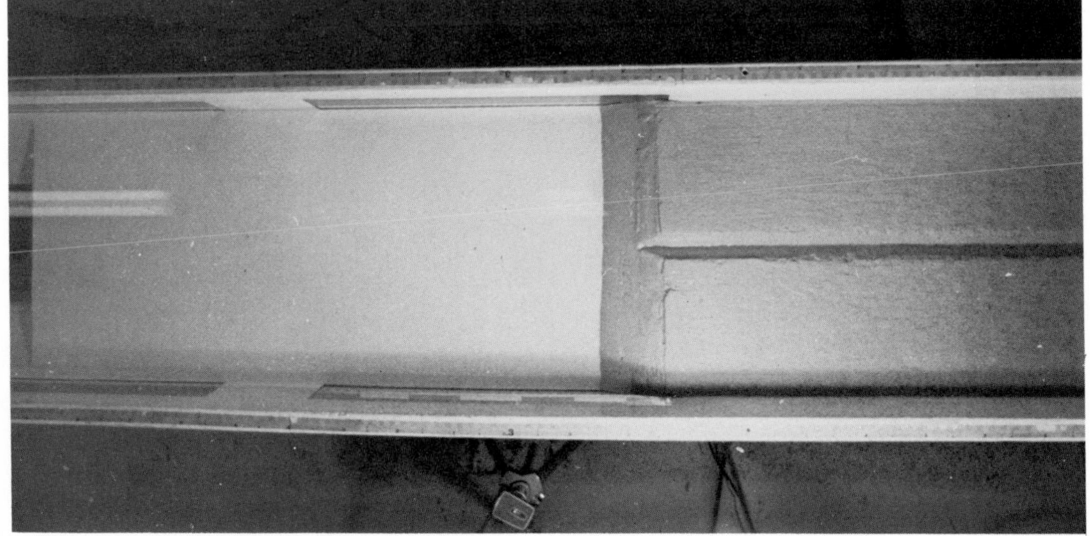

Figure 8.81. Experiment 10 at minute 5.5: commencement of delta.

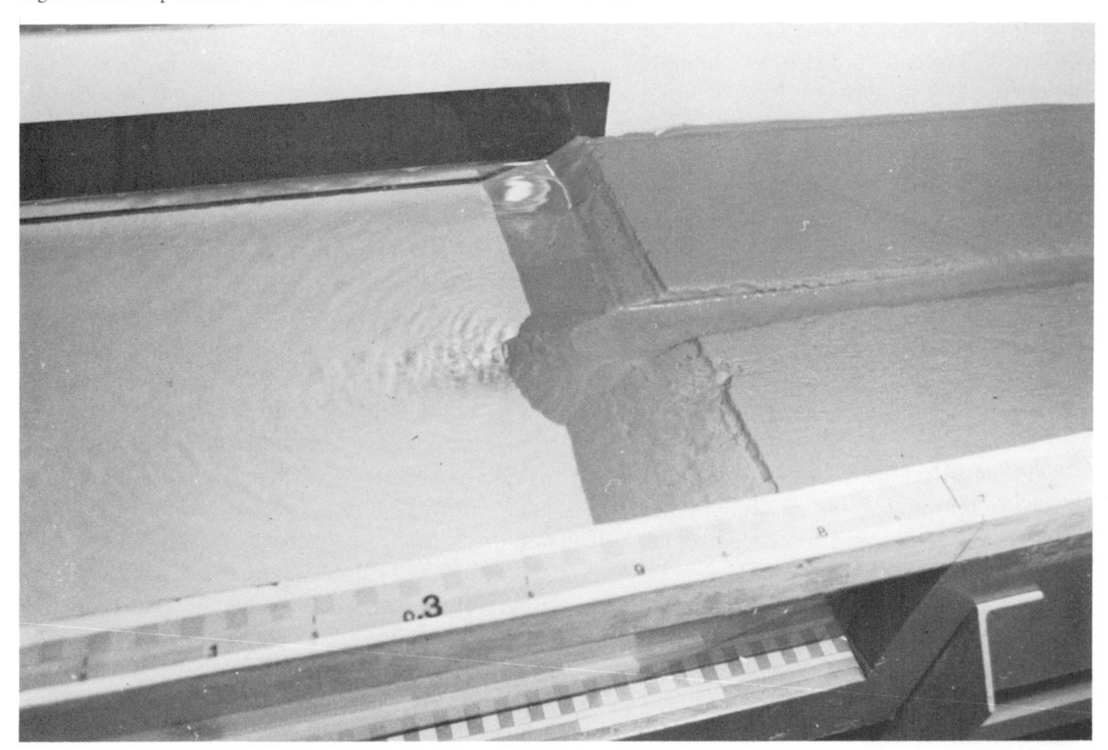

Figure 8.82. Experiment 10 at minute 21, showing submarine deposition on delta.

original submarine slope (Figure 8.82). By MINUTE 25 a small quantity of fines was deposited a few centimetres beyond the foot of the delta (Figure 8.83). Material was added to the delta in a series of pulsing jets at 3 or 4 second intervals. Velocity of flow in the stream was fairly constant at 28.6 cm s^{-1}, with turbulent flow in the lower reaches; sea-current velocities were: first 50 cm from cliff, 17.9 cm s^{-1}; and next 50 cm seaward, 6.4 cm s^{-1}. The delta form at switch-off (MINUTE 43.3) was semi-circular, extending 12 cm from the cliff foot, 13.3 cm in width, with some eight rudimentary distribution lobes. A 6.0 cm wide bed trail of fine sand extended 30.0 cm ahead of the delta toe and a dust trail extended another 80 cm (Figure 8.84).

Conclusions: The stream eroded a rectilinear channel which did not reach equilibrium during the duration of the experiment. Dislodged material was carried, by traction and suspension by turbulent flow, onto and out across the delta plain and onto the delta slope, which maintained the same angle of rest as the coastal zone. The deposition of topset beds beneath the delta plain, and foreset beds beneath the delta slope, was affected in a narrow sector by a lobe of material which periodically became too gentle in slope for efficiency, forcing the water to flow to either side to form a new lobe. A secondary accretion mode was by a jet of suspended material that shot off the delta every three or four seconds to fall in a diffuse mass on the delta slope, or beyond in the prodelta zone. Fines were carried further out to the sea-bed. The total delta complex was therefore tri-zonal but with minimal development of the deep water deposits, due to lack of fines.

Systems interpretation: The delta system has an input from the stream of water and sediment. Since the delta extends from about 2 mm below sea-level to the sea bed (~4.8 cm) there is a height value for potential energy. The delta mass was 203.3 g and its volume 133 cm^3. Output from the delta system was primarily in the form of water moving away from the depositional layers. The delta was continually growing, and thus its mass and volume increasing, but at a diminishing rate as the stream became more graded and its erosion rate decreased. The system thus comprised a morphological subsystem containing some constant

Figure 8.83. Experiment 10 at minute 25, with fines deposited ahead of main delta mass.

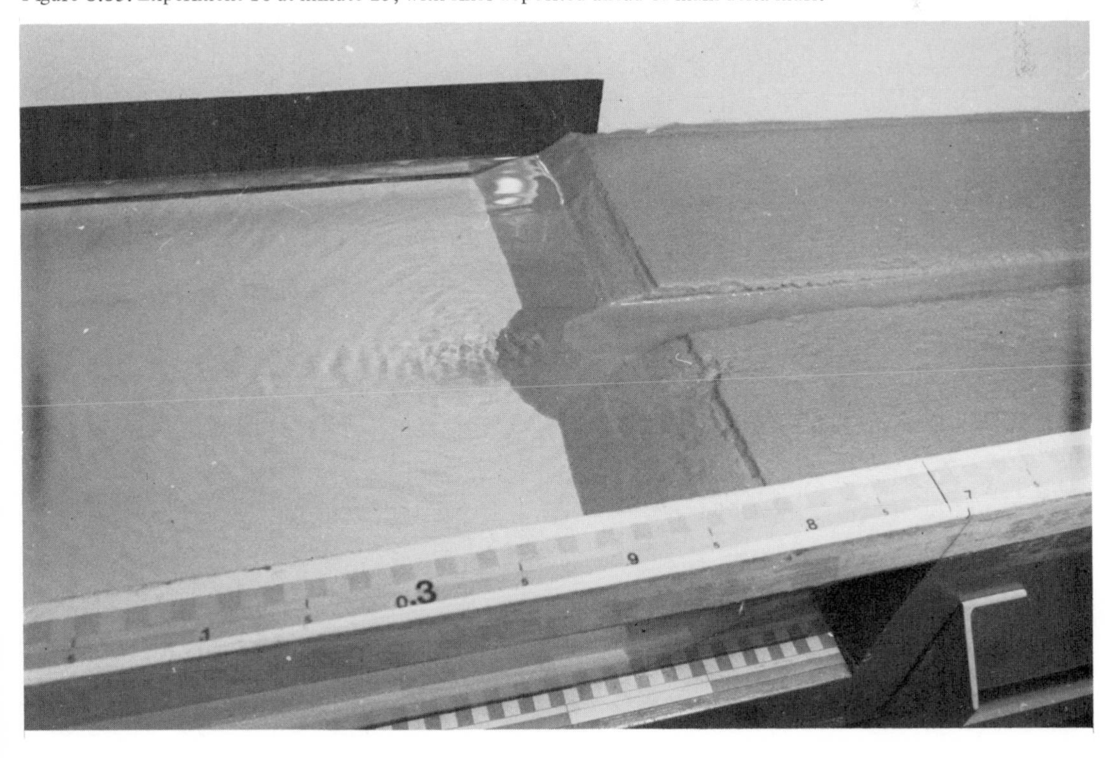

parameters, such as delta slope and circularity of plan, while extending laterally but not vertically, giving some equilibrium factors, but as a total form being in inequilibrium.

Although delta mass increased continually, the cascading subsystem decreased, for although the water mass and energy remained constant, the sediment mass diminished while the height of the delta remained constant. It may also be reasoned that, if the water component of the cascade is essentially that of water moving on from the zone of deposition and consolidation, the water component must also decrease, with decrease in deposited material. The delta is clearly a system closely linked to river dynamics, which it reflects, while surrounded by the marine or lake system.

Experiment 11

Purpose: To produce deep-water sedimentation.
Apparatus: Flume; small chute 1 m long × 3 cm wide × 3 cm deep resting on an artificial cliff, preferably of clear plastic (Figure 8.85); feed tube extension of constant-head feed; sand hopper with vibrator mechanism; three grades of sand, coarse, medium and fine.
Procedure: (Note that the artificial, fixed bed, allows a higher stream velocity than a mobile bed, with greater seaward transportation potential than, say, Experiment 10.) The coarse and medium sands were washed, sieved and dried. The three grades were weighed into slugs (see controls). The hopper was set up and the feed rate adjusted. The discharge was pre-set. The small chute was set up in the head of the flume, giving about 2.0 m of sea for deposition. The main flume was filled with water until the water was half-way along the bed of the sloping chute. The discharge was commenced by unplugging the pre-set pipe. Sand was fed in the sequence

coarse → medium + water flush → (sea change) → fine → medium + water flush → (sea change) → coarse

Figure 8.84. Experiment 10 at minute 43.3 (switch-off).

Notes were taken on the seaward and rear side of the chute support (cliff) to determine superficial and sectional characteristics of deposition.
Controls: Discharge constant at 47.7 $cm^3 s^{-1}$.

Sand grades	Through (mm)	Caught (mm)	Slugs (g)
Coarse	1.0	0.5	750 (2)
Medium	0.5	0.177	880 (2)
Fine	0.177	0.0625	525 (1)
Total			3785

Duration of sediment feed: coarse 11.63 minutes; medium 7.33 minutes; fine 8.18 minutes; medium 7.37 minutes; coarse 9.7 minutes. Approximate sand feed rate: coarse 0.89 $g s^{-1}$; medium 1.77 $g s^{-1}$; fine 1.88 $g s^{-1}$. Sea-level constant at 12.4 cm; distance of water flow from input to the sea 60.0 cm; height of feed point from sea-level 3.5 cm; slope of chute 1 in 40; water velocity in chute 79 $cm s^{-1}$; water velocity in sea 26 $cm s^{-1}$, first 50 cm offshore; water velocity in sea 8 $cm s^{-1}$, next 50 cm offshore.
Observations and data: During the introduction of 750 g of coarse sand the velocity of particles in traction decreased rapidly at the stream mouth, and being too large to be carried beyond the cliff face they dropped to form an arcuate delta 6.0 cm high by 21.5 cm wide and extending 11.0 cm into the sea. No deep-sea deposits occurred beyond the delta (Figure 8.86), demonstrating that bottomset and deep-water material would have to be of lesser size than 0.5 mm.

The next slug, 880 g of medium sand, contained particles small enough to be carried by saltation and suspension and, therefore, to remain water-borne until beyond the delta. As the stream water and sea-water were of equal density, flow by axial jet occurred and a bottomset bed was laid down, extending 30 cm beyond the delta foot and 14 cm wide. Out to sea this deposit was horizontal and uniformly thin and, geomorphically, did not form part of the delta (Figure 8.87). Introduction

Figure 8.85. Small flume and hopper equipment for Experiment 11.

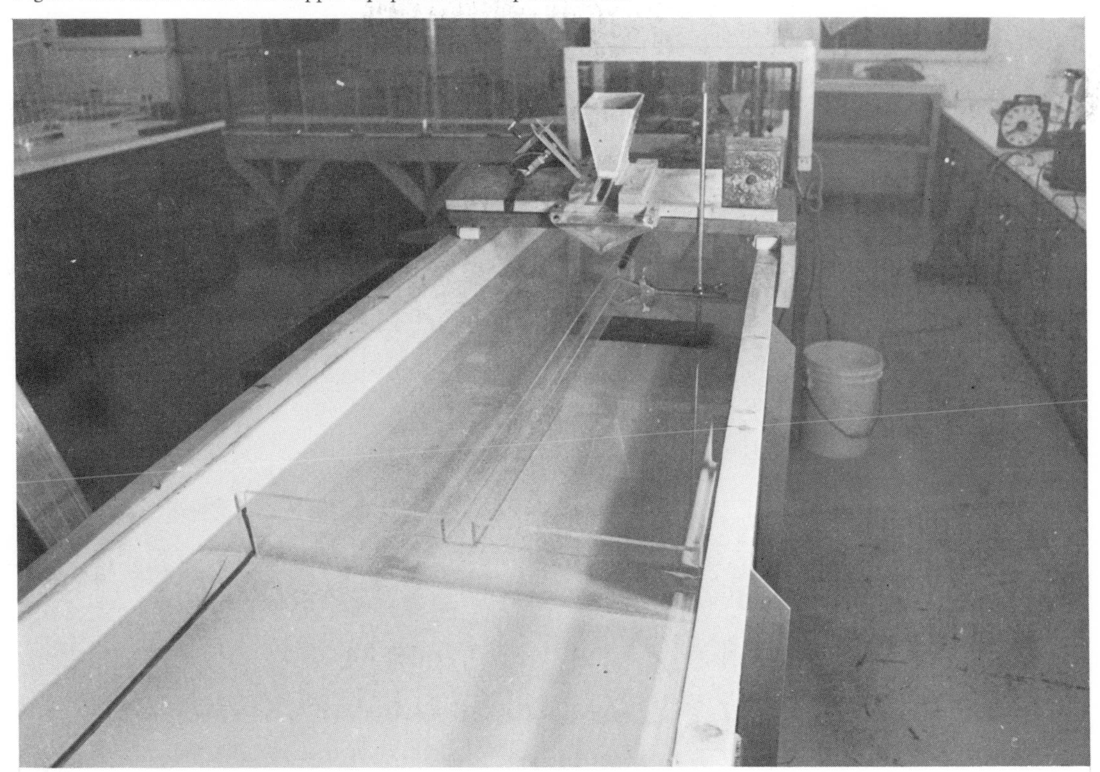

Figure 8.86. Experiment 11 after introduction of coarse sand, showing absence of fines beyond the delta.

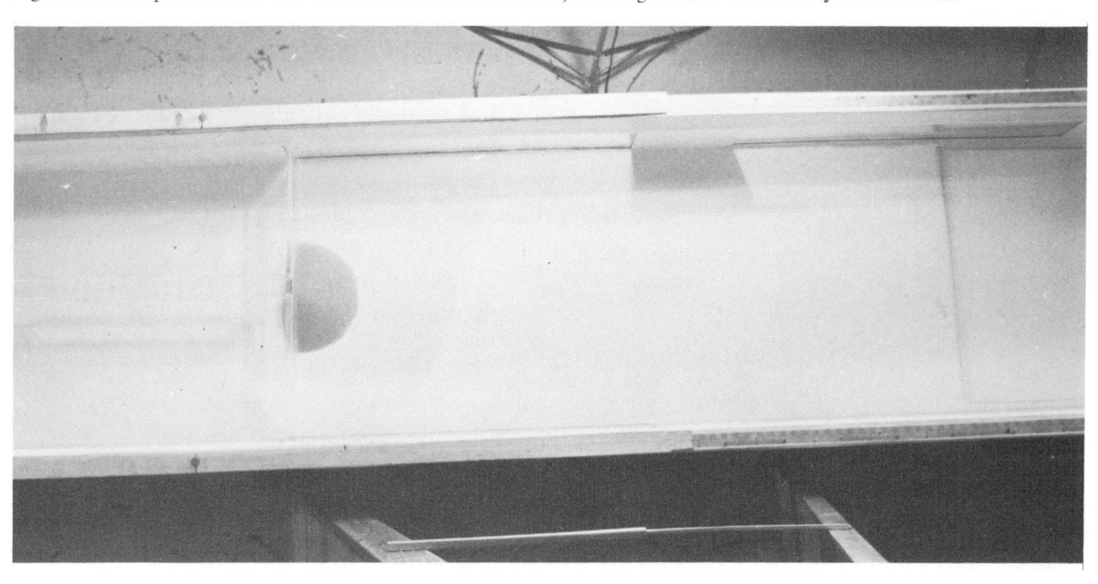

Figure 8.87. Experiment 11 after introduction of medium sand.

of fine sand, inclusive of dust, permitted a transportation mode almost entirely of suspension, and the axial jet carried material right over the delta, leaving only a small proportion behind. Within the 8.18 minutes of this phase the sea became occluded and dirty water began to pass through the outlet pipe. The sea area acquired a thin deposit of fines with a concentration of coarser material towards the source (Figure 8.88) forming a prodelta tongue 22.0 cm long, 7.0 cm wide and 2.0 cm deep at the delta foot junction.

The second slug of medium material, over 7.37 minutes, was again primarily of deltaic grades forming a uniform layer on the delta slope (Figure 8.89) and building up the prodelta by deposition upon the preceding fines (Figure 8.90).

The final 750 g slug of coarse sand settled firmly on the delta core, contributing neither to the prodelta nor to the deep sea deposits (Figure 8.91). The post-run analysis of sediment distribution is given in Table 8.1. The final delta form was symmetrical and contained 3151 g of sediment, while the deep-water deposits contained 592 g.

Conclusion: Deep-water deposits in this experiment are produced by a combination of controls, namely: a stream with a velocity adequate to continue out to sea; presence of sediment fine enough to be carried in suspension; and sufficient time to permit settling of suspended material. Of these three

Table 8.1. *Post-run analysis of sediment distribution*

Grade (mm)	Input (g)	Delta (g)	Sea bed (g)	Short-fall (g)
1–0.5	1500	1451	0	49
0.5–0.177	1760	1510	248.5	1.5
0.177–0.0625	525	188.8	333.5	2.7
Dust		1.5	10.2	(−11.7)
Totals	3785	3151.3	592.2	41.5

Figure 8.88. Experiment 11 after introduction of fine sand, with widespread deposition of fines.

Figure 8.89. Experiment 11 after introduction of second slug of medium sand, showing formation of uniform layer on delta slope (section to rear through plexiglass plate).

Figure 8.90. Experiment 11 after introduction of second slug of medium sand, showing build-up of prodelta.

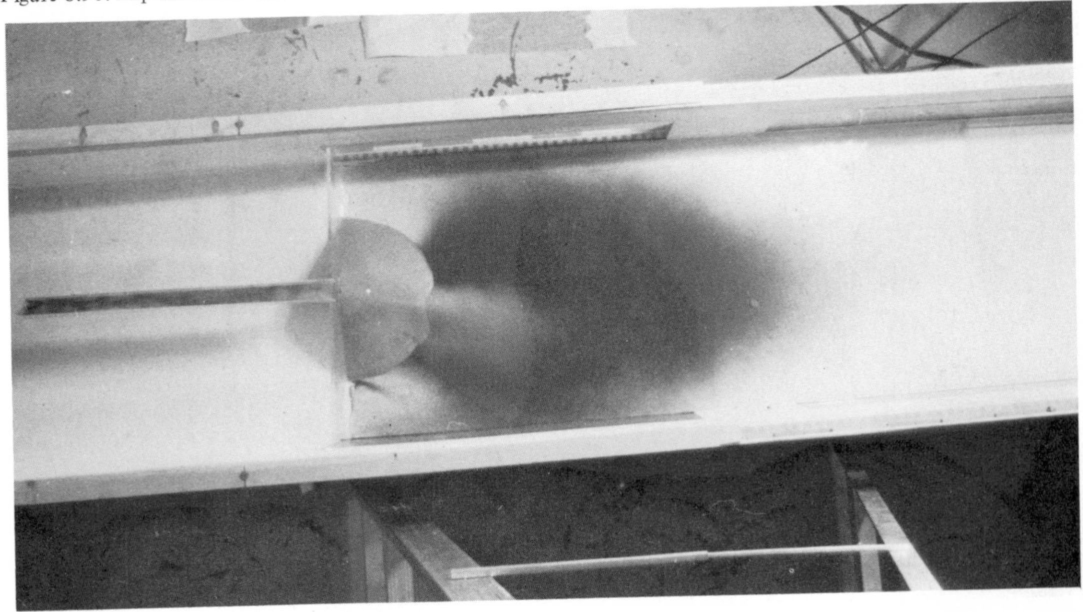

factors the third was barely adequate since turbid water entered the flume's exit pipe.

Owing to the modal distribution of sand grades the sharp distinction between deep-water and deltaic deposition was demonstrated, with sand coarser than 0.5 mm falling entirely on the delta, sand between 0.5 and 0.177 mm phasing from traction to suspension in transport mode, and falling 36% on the delta and 64% on the sea bed, while material of less than 0.177 mm was deposited entirely beyond the delta slope, much of it on the sea bed.

Systems interpretation: The deep-water deposit as a system is ill-defined since the outer boundary phases out horizontally into ever thinner deposits over the sea bed (Figure 8.92). Furthermore, with the input occurring by settling from turbid water, the vertical boundary is similarly indefinite due to diffusion phenomena at the water/bed interface, and lack of bed-compaction. Consequently, identification of the system is to a considerable extent arbitrary and it is only after the removal of the sea that a distinct vertical boundary occurs - but this is not in the dynamic solid-water context. The lateral boundaries are at the flume walls.

Given these obscuring aspects, the bed deposits can be seen to form part of a system which develops with an input of particle sizes of less than 0.5 mm and sea-current velocities of less than $8\ \mathrm{cm\ s^{-1}}$ (it being assumed that the minimal surface velocity was greater than the velocities at depth). The deposits covered about $1.68\ \mathrm{m}^2$ and had a mass of 592.2 g. Bed thickness varied considerably but ranged from ~0.0–20 mm. If the sea above the deposits is regarded as being a separate system, there appears to be no output from the bed deposit at the early stages, so that a condition of static equilibrium obtains. Increments of material merely produce an increase in bed-deposit dimensions. It will follow, however, that, as thickness increases, loading of the original bed (trough or crust) could result in distortion and failure, as in a geosyncline or other movement with a gravity component, with instability or equilibrium according to the effectiveness of adjustment. At this stage the potential energy of the accumulated mass would become converted, possibly catastrophically, to kinetic energy and mass might move from the system. Such a condition is simulated in part in the pressure box experiments of Chapter 11.

Experiment 12

Purpose: To produce pools and riffles.

Apparatus: Flume; medium sand; curved carriage blade; timer; glassware.

Procedure: A test bed of clean uniform sand was made up in the flume, measuring 2.8 m in length and terminating at the edge of the sump to avoid

Figure 8.91. Experiment 11 after introduction of final slug of coarse sand.

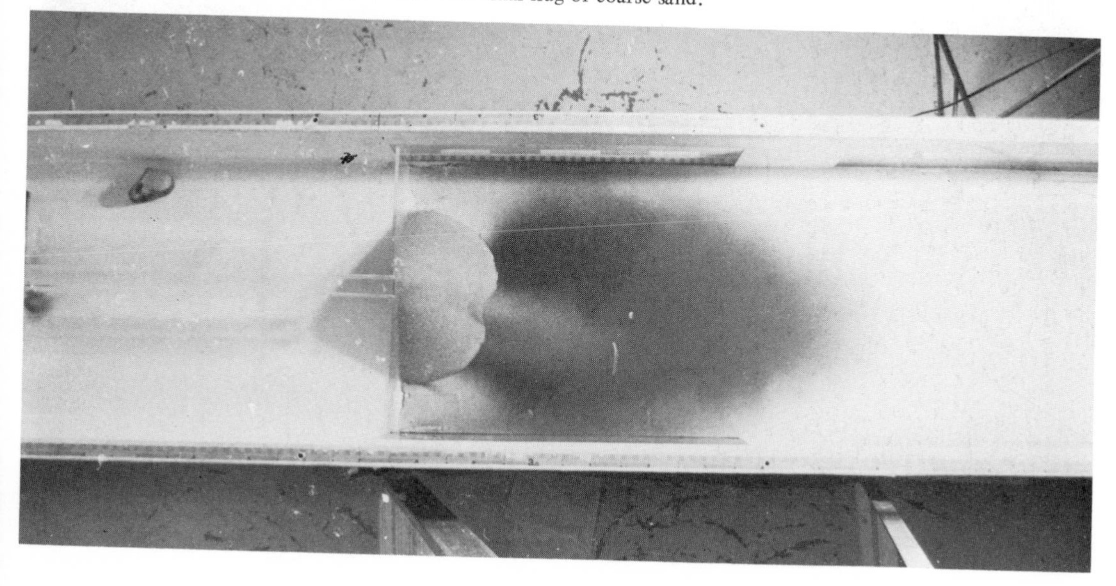

Figure 8.92. Post-run photo showing ill-defined boundary of deep-water fines.

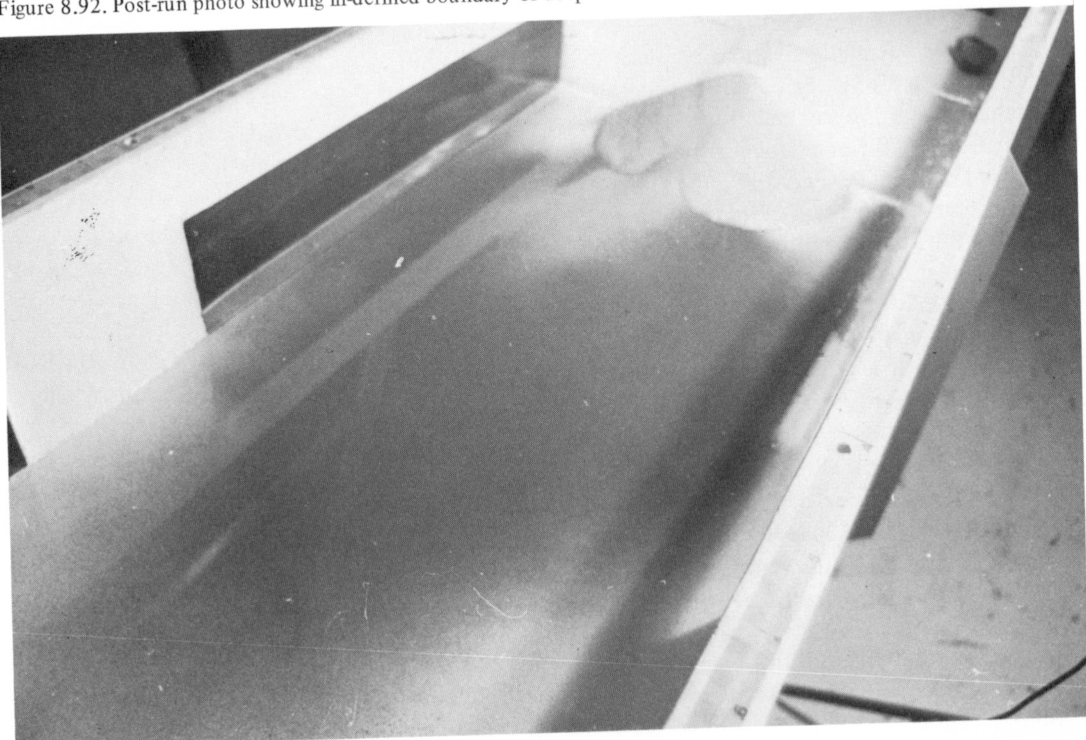

Figure 8.93. Test bed for Experiment 12, showing regular seaward slope.

delta formation. The upper surface was moulded by a curved blade of about 1.0 m radius attached to the carriage which ran on two side wedges to give a constant seaward slope (Figure 8.93). The sand surface was trimmed and brushed smooth. Sea-water was introduced and the seaward edge of the test bed was cleaned up, after which the sea was removed, the sump cleaned up and the sea re-introduced. The water supply was pre-set and the headpool allowed to rise to operating level.
Controls: Discharge 61 cm s^{-1}; bed slope 1:33.3; sea-level constant at 2.9 cm; headpool 12.2 cm deep at commencement falling to 11.3 cm deep at termination; stream velocity 48 cm s^{-1} (fairly constant); duration 74.87 minutes. No sediment added to system.
Observations and data: Flow strongly turbulent. At MINUTE 4 channel forms and incipient pools were seen at about 100 cm intervals on-centre (i.e. metre 1.5, 2.4 and 3.5) as shown in the composite photograph (Figure 8.94). At MINUTE 13 the pools were in approximately the same positions with a fourth pool characteristic at the mouth (metre 3.8). Water extended across a main channel of 20-30 cm in width but was incising a 3.5 cm wide channel in sections downstream from the pools. By MINUTE 19 the pools had become incised channels at metres 1.62, 2.55 and 3.62 as shown in the composite photograph (Figure 8.95). By MINUTE 30 the pools were regularly spaced at metres 1.6-2.1; 2.56-2.78; 3.49-3.56. By MINUTE 35 a distinct channel existed from the headpool to metre 2.4, of 7.0 cm width and 3.0 mm depth. The velocity in this channel was 68 cm s^{-1} and the slope of the water surface was 1 in 40. At MINUTE 50 a single long curved channel persisted (Figure 8.96). By MINUTE 60 there was a deep regular channel forming new pools at metres 1.65, 2.15, and 3.7. At MINUTE 74.87, switch-off. The single curve channel had persisted (Figure 8.97) and pools were seen to occur as shown in Table 8.2.

The final overall erosion form of the model

Figure 8.94. Experiment 12 at minute 4 showing incipient pool development. (The distorted perspective in this photograph is due to it being a montage of two photographs, as in Figure 8.95.)

Figure 8.95. Experiment 12 at minute 19 (montage), showing incision of pools.

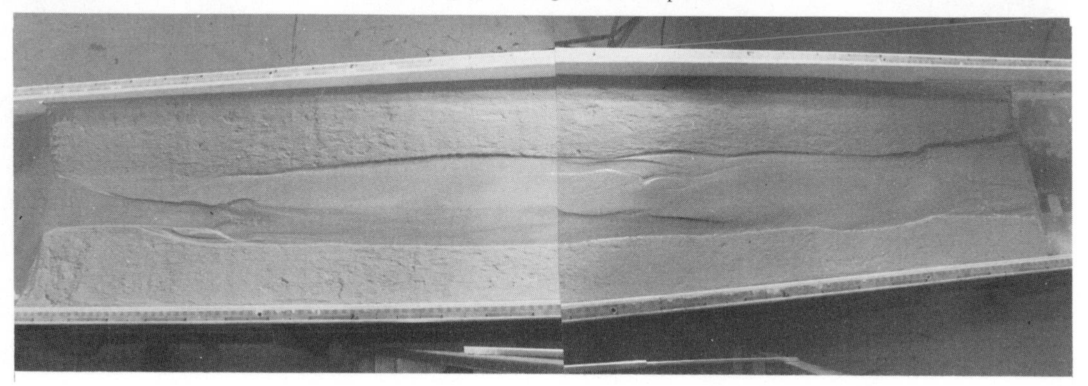

Table 8.2

		Spacing of pools			
Metre	Length of pool (cm)	Centre point (m)	Distance	× 7.083 to give space	Channel width (cm)
1.25–1.43	18	1.34	0.29	4.0	
1.56–1.70 major	14	1.63	0.85	12.0	7.0
2.40–2.57 major	17	2.48	0.78	11.0	11.5
3.20–3.32 major	12	3.26	0.68	9.6	23.5
3.92–lip of sump	5	3.94			

Figure 8.96. Experiment 12 at minute 50, with a single, long, curved channel.

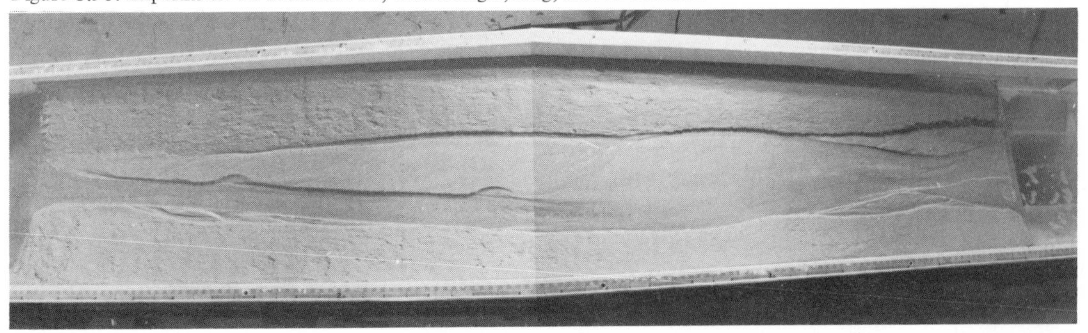

Figure 8.97. Experiment 12 at minute 74.87 (switch-off), showing general channel conformation.

consisted of a broad valley incised into the initial surface, varying in width from 8.0 cm at the head to 33.7 cm at the mouth (Figure 8.98). Within this an inner channel occurred, varying in width from 7 cm near the head to 23.5 cm at metre 3.0 and 33.7 cm at the mouth. Pools and riffles occurred within this inner channel, as shown in Table 8.2 and Figure 8.98, with pool depth exceeding mean channel depth of 3 mm and riffle depth less than 3 mm.

Conclusions: The flow of water at the commencement of the experiment immediately concentrated into a pathway, a channel of about 11.0 cm was formed and the first pools developed, more or less equally spaced, at 1.0 m intervals on alternate sides of the channel. Pools with interspaced riffles were present throughout the experiment; they occasionally shifted their position and sometimes appeared to be irregularly spaced, but would reform later at a more regular spacing. When spacing was irregular, long spaces were perceived to be multiples of short, as at MINUTE 60, when the recorded spaces were 155 cm and 50 cm. There was an intimate relationship between the channel centre-line and pools of the type noted at MINUTE 35, when the broader initial channel suddenly gave way to a narrower (7.0 cm wide) channel which seemed perfectly to take up the single long curve of the wider channel but to have a distinct edge. This narrower channel existed in segments at pools throughout the experiment while the broader, less defined reaches were associated with riffles. The whole system of broad and narrow channel, riffle and pool formation and spacing was inter-related in a dynamic way. The final spacing at MINUTE 74, as given in Table 8.2, demonstrates considerable irregularity at that point in time. Inspection shows that, of possible factors that would divide fairly accurately into these spaces, 7.083 gives the best results, for which products are 4.0, 12.0, 11.0 and 9.6, suggesting that there is a hydraulic relationship between a fundamental channel width of about 7 cm and a fundamental pool spacing of about 7 cm. Since pools were usually located on alternate sides of the mean flow path, a rudimentary curvature in the centre-line occurs, further suggesting a relationship between stream sinuosity and pool spacing.

Systems interpretation: The complete channel may be taken as a system, in which case the morphological subsystem was complex, incorporating a broad (20 cm) initial channel, which later functioned as a containing valley, and a secondary 7 cm wide channel which was more minutely involved with system dynamics. The cascading subsystem, at first related to the broad channel, later established the narrower channel

Figure 8.98. Schematic diagram of final channel conformation.

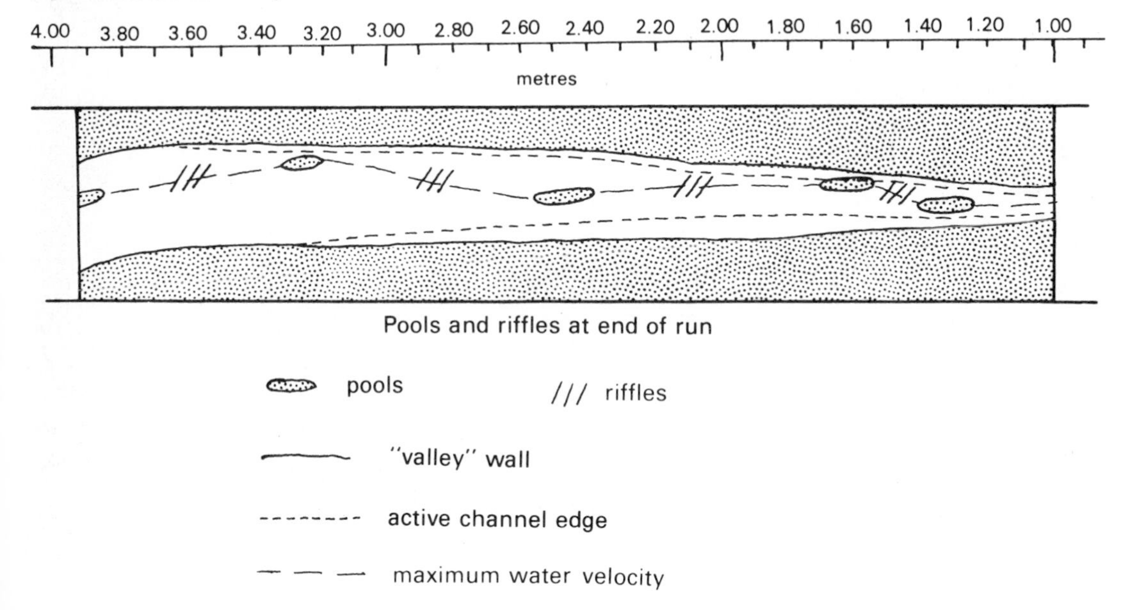

and persistently re-established it as the preferred morphological subsystem. The cascading mass was 274 024 g of water and 13 970 g of sediment, with a fall of 8.4 cm, to give energy values of $E_p = 225.7$ J (water energy) and 11.5 J (sediment energy) during the 74.87 minute run. This mass and energy throughput, in its close association with the narrow channel, developed pools and interspaced riffles at spaces in multiples of 7.0 cm, i.e, multiples of channel width.

The broader valley slowly widened towards the mouth throughout the run as the narrower channel shifted from side to side, and may be regarded as being in inequilibrium even in the short duration of the experiment, while the smaller channel, which constantly broadened and shallowed and then reverted to its preferred narrow, pool-containing form, was in a condition of steady-state equilibrium. However, towards the end of the experiment, perhaps with slow change of gradient, even the narrow channel subsystem became less definite, indicating that the subsystem might then be better regarded as being for a time in dynamic equilibrium, and then with ultimate loss of channel form and random distribution of sediment and water, approaching a condition of entropy.

Experiment 13

Purpose: To study sedimentation in a curved channel.

Apparatus: Plexiglass channel with a curved centre section with radii 30 cm and 40 cm and straight end sections, measuring 1.76 m overall by 10 cm inside width by 5 cm depth (Figure 8.99); vibrating sand hopper; controlled water-feed system (main flume); clamps; timer; glassware, etc.; three grades of sand.

Procedure: The small plexiglass channel was set up in the main flume as shown in Figure 8.99. The channel slope was set at 1 in 60 and a vibrating

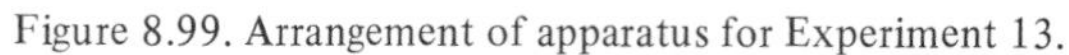

Figure 8.99. Arrangement of apparatus for Experiment 13.

hopper was located at a point 3.0 cm down from the channel head but not in contact with the channel (Figure 8.100). The water input was pre-set from the main flume using a free-fall linkage with a plastic rain gutter into the small channel head, and upstream from the sand feed. Dry, clean sand was thoroughly mixed and placed in the hopper as a single charge. The hopper was pre-set to give the required feed rate. A catching dish was placed beneath the mouth of the small channel to collect the solids output.

At the start of the run, water input was commenced, then sand feed, and timing began with the first sand striking the channel. The experiment concluded when the hopper was empty, and, during post-run data collection, sediment was removed from the channel in three zones with a scoop of one-third of the channel width.
Controls: Water input 177 $cm^3 s^{-1}$; sand input 1.9 $g s^{-1}$; sand mix as in table.

Grade (mm)	Weight (g)
1.0–0.5	1000
0.5–0.177	1000
0.177–0.0625	500
Total	2500

Table 8.3. *Distribution of sand grades in the channel*

Grades (mm)	Zones: Inner (g)	Middle (g)	Outer (g)	Totals (g)
1.0–0.5	118.8	136.5	79.9	335.2
0.5–0.177	158.4	181.5	91.5	431.4
0.177–0.0625	59.5	82.1	54.6	196.2
Fines	1.0	0.9	0.6	2.5
Totals	337.7	401	226.6	965.3

Channel slope 1:60; duration of run 22 minutes.
Observation and data: The water flow was turbulent, and at MINUTE 22 the water level was observed to stand higher on the outside than the inside of the bends, as shown in Figures 8.101 and 8.102, the latter of which covers the sector A in the diagram. Sand was deposited in some parts of the bed while others remained sediment-free, and differentiation occurred into zones of coarse, medium and fine grades, appearing to obey no obvious rule (Figure 8.103). Dry, sieved sediments from the equal-area outer, middle and inner zones gave the proportions as shown in Table 8.3.

These data show that, of all sediment added to the system, 38.6% was deposited in the curve

Figure 8.100. Experiment 13 before commencement. Showing hopper out-of-contact with flume.

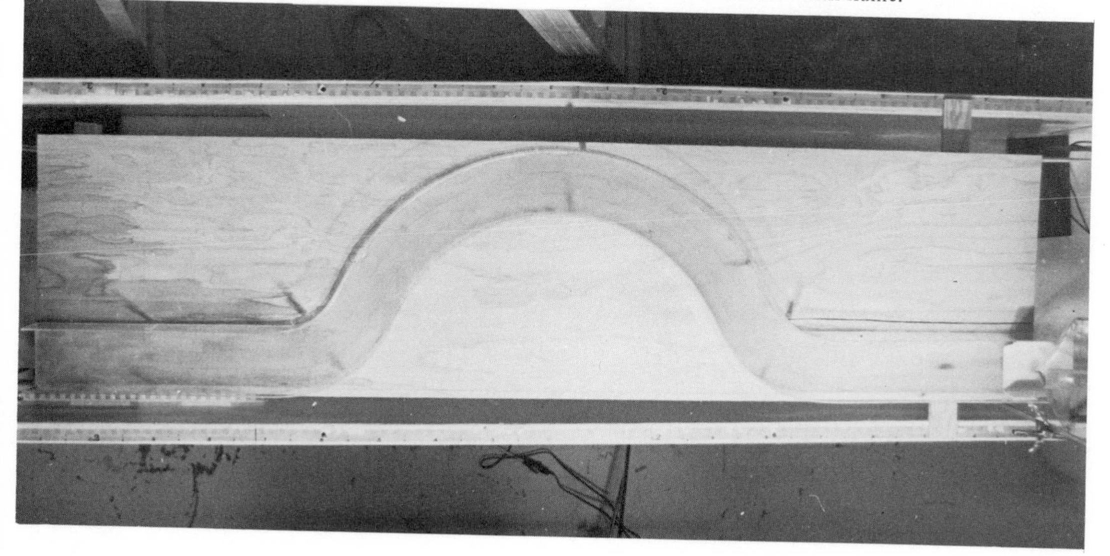

Figure 8.101. Schematic diagram for conformation of water surface at minute 2 (Experiment 13).

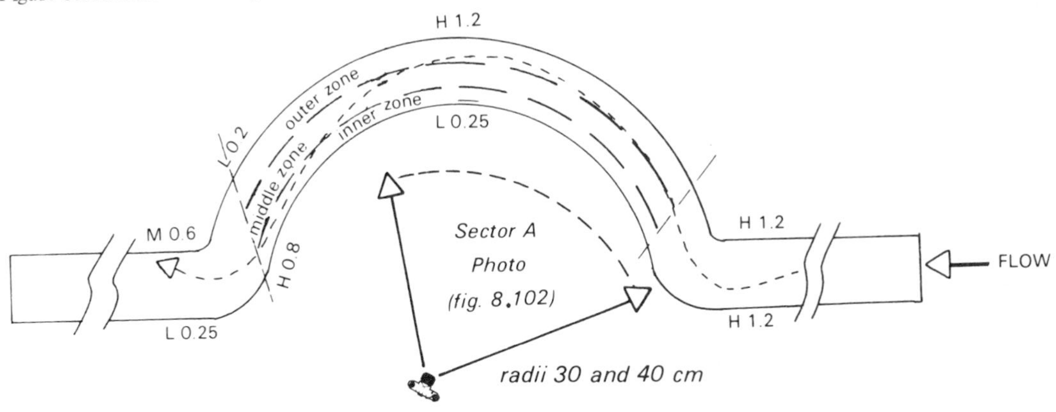

water surface in curved channel (locations in cm at minute 2)

◁- - - - - - - - - line of maximum velocity

H = high water surface in centimeters

M = medium water surface in centimeters

L = Low water surface in centimeters
(may be water-free bed)

Figure 8.102. Experiment 13 at minute 2, showing differential water levels in curves.

during the 22 minute duration. Of the coarse material added, 33.5% was deposited in the curved section; of the medium material 43.1%, and of the fine material, 39.2%. Within the three curve zones, the inner zone had 35.0% of the total curve deposits, the middle zone had 41.5% and the outer zone had 23.5%. The inner zone deposits were: 35.2% coarse; 46.9% medium; 17.6% fine. The middle zone deposits were: 34.0% coarse; 45.3% medium; 20.5% fine. The outer zone deposits were: 35.3% coarse; 40.4% medium; 24.1% fine.

Velocity readings were taken after the experiment in the three zones by dye and stop-watch, but could not be differentiated with confidence.

Sediment output from the system during the run totalled 443.3 g dry weight (17.7% of input), made up of: coarse sand 301.9 g; medium sand 112.7 g; fine sand 28.0 g; fines 0.7 g.

Conclusion: Water entering the curved segment rose high on the concave side and fell away from the convex sides by reason of centrifugal force, giving a mean difference in cross-channel water depth on the three bends of 212%. Sediment transported by the water was carried into and out of the bend, but 38.6% remained as a deposit within the bend. Inspection of the surface and side exposures (through the plexiglass) gave an imperfect idea of sediment grade distributions, but the inner zone received 35% of the deposits while the outer zone retained 23.5%. Of the three zones, the middle was the zone of optimum deposition with 41.5%, which may be seen as a cross-over zone characteristic and perhaps related to processes of bar development (Figure 8.104). The fractional separation of sediment grades in each zone showed that coarse material was deposited in about the same percentage: inner zone (coarse) 35.2% of total deposit in the zone; middle zone (coarse) 34.0% of total deposit in the zone; outer zone (coarse) 35.3% of total deposit in the zone.

The medium grade fraction was distributed: inner zone (medium) 46.9% of total deposit in the zone; middle zone (medium) 45.3% of total deposit in the zone; outer zone (medium) 40.4% of total deposit in the zone.

The fine sand was distributed: inner zone (fine) 17.6% of total deposit in the zone; middle zone (fine) 20.5% of total deposit in the zone; outer zone (fine) 24.1% of total deposit in the zone.

These data show no substantial difference in coarse material deposition, comparatively, between the zones, which may be explained in terms of transportation by traction, with low velocity-differences in the bed layer. Medium grade sand

Figure 8.103. Experiment 13 at minute 22, momentarily before switch-off, showing distribution of sand in plan view.

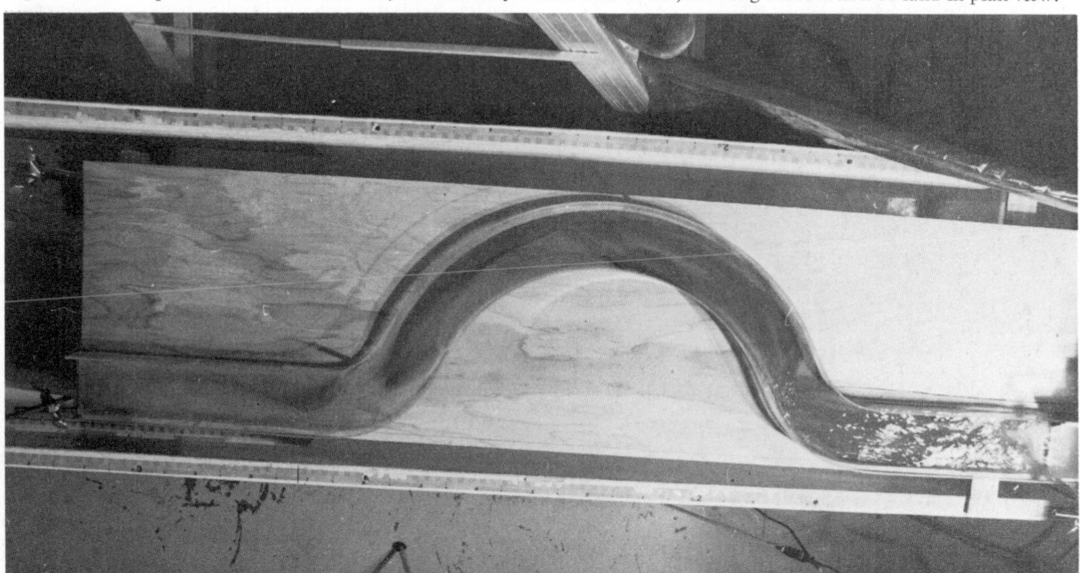

shows a 6.5% difference between high proportions in the inner zone, where lower velocities occur, and the outer, higher-velocity zone. Since medium grades will move partly by traction but more often by saltation and suspension, this result accords with general theory. Fine grades exhibit the same difference of 6.5%, suggesting that there was no substantial difference in transport mode between fine and medium particle sizes. The lower absolute values of fine sand in these results reflect, in part, the smaller quantity of these grades in the original sample (500 g as distinct from 1000 g for coarse and medium).

Figure 8.104. Experiment 13 after cessation of water-flow, showing a cross-over zone.

Systems interpretation: The single central curved segment is regarded as a system, and the straight segments and reverse curves are excluded. The elementary channel form is, therefore, an arc of a circle 70 cm across, 10 cm wide and 1.5 cm deep (greatest water depth). This becomes modified by deposition into an irregular bed form, while retaining roughly the same lateral boundaries.

Mean water velocity was $0.413\ \mathrm{m\,s^{-1}}$ and, as discharge was $177\ \mathrm{cm^3\,s^{-1}}$, the mean cross-section was $4.29\ \mathrm{cm^2}$. As the width was 10 cm, the mean depth was 4.29 mm, and the effective volume of the channel was $472\ \mathrm{cm^3}$, representing the morphological subsystem. The mass of water passing through the system was 233 640 g in 22 minutes, and the mass of sand passing through was 2500 − 965.3 (remaining in curved section) − 546 (in the upstream end of the channel that did not enter the main curve) = 988 g. Drop distance is 1.17 cm in the system curve. The energy of the system is therefore:

water - $(233\,640 \times 1.17)/10\,197.16 = 26.8$ J;
sediment - [(sediment passing through $+\frac{1}{2}$ sand in system) × 1.17]/10 197.16
$= [(988 \times 1.17) + \frac{1}{2}(965.3 \times 1.17)]/10\,197.16$
$= (1156 + 564.7)/10\,197.16 = 0.17$ J.

During the run an equilibrium channel bed (section) developed with sediment entering to replace sediment leaving, maintaining a stable condition and representing steady-state equilibrium, the changes being small and momentary. This was a condition caused by an immobile plexiglass channel permitting no vertical erosion or bed grading other than by deposition, and the sediments conformed to steady hydrodynamic influences to produce an integrated form. If a very much longer time scale were envisaged, in which the plexiglass was eroded, the system would have entered a phase of dynamic equilibrium.

9 The wave tank

The production of waves to impinge upon a sandy coastline is perhaps the commonest of models used in instructional situations and references to the technique are numerous. Balchin and Richards[1] suggest an out-of-doors tank with a wave generator of Meccano operated by hand. Use of a stream table with an electrically driven wave generator is suggested by Schwartz.[2]

A comprehensive review of the literature and theory of hardware modelling as concerned with wave generation is given by King,[3,4] while Bascom[5] and Vallentine[6] treat the mathematical aspects of wave mechanics in a clear manner. For introductory university work, Ritter's[7] *Process geomorphology* suggests a reasonable standard of theoretical treatment and nomenclature.

The use of waves in models lends itself to more precision than is commonly acknowledged in demonstration literature. In the school or college laboratory waves may be generated to act upon shelving, sandy coastlines by observing basic mechanical factors.

Figures 9.1 to 9.4 show a number of simple wave generators, ranging from a weak, slap-the-water commercial model, which is of little value except for the most elementary and uncontrolled experiments, to a versatile device attached to a flume and based upon specifications published by C. L. Stong.[8]

The most convenient means of displacing water for demonstration of shore processes is a paddle which moves back and forth, pushing and pulling alternately to generate waves in front of and behind the paddle. The front waves travel on to impinge upon the model beach, but the rear waves are not required and are damped out by placing fibre or mesh baffles behind the paddle. A versatile generator, as used for experiments in this chapter but with a narrow paddle, is shown in Figure 9.5, and comprises a $\frac{1}{4}$ h.p. electric motor geared down and incorporating a variable speed control. The rotational motion of the drive shaft is transformed to a back and forth motion by an eccentric, which adjusts to vary the stroke. The paddle is hinged at the top, but a bottom hinge is probably more common.

The construction should be robust: as M. A. Morgan[9] points out. 'There is no substitute for really rugged construction and accurately designed wave generating equipment . . . one tends to forget the great weight of relatively small quantities of water.'

The paddle should sweep within a millimetre or two of the sides and floor of the tank – that is to say, it must be designed to fit the tank.

Figure 9.1. A small commercial water-slapping wave generator.

The baffles can be of plexiglass or waterproof plywood, drilled with a honeycomb of 2.5 cm holes and backed by fibre mat set at an angle of about 30°. This should be placed as far back from the rear of the generator as is convenient. Baffles and generator should be firmly fixed to the tank to prevent shifting during an experiment.

The present experiments made use of an existing 6 m flume, but a wave tank for beach-profile study can be much narrower, say 10-20 cm, and for the present experiments need only have been 3.0 m in length.

Figure 9.2. A simple rotating-paddle wave generator.

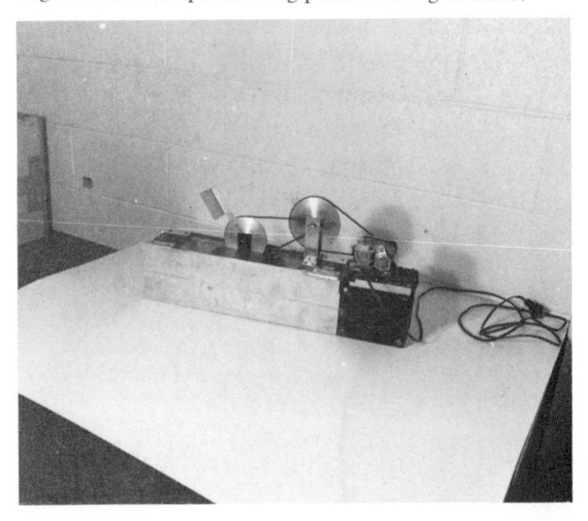

Figure 9.3. A to-and-fro action adjustable-pitch wave generator.

The beach material used was the standard medium sand employed in most experiments, which may therefore be shaped into a beach by setting up, wetting, and shocking to consolidate, and then cleaned with scrapers and brushes.

The important wave characteristics are wave frequency or period, wave length and wave height, and it is a minimum generator requirement that these may be varied at will.

The number of waves per unit of time (usually seconds) or, more commonly, the number of seconds per wave is controlled by the number of seconds between forward pushes of the paddle. This will also control wave length since, for a sinusoidal wave $L = (g/2\pi)T^2$, or $L = 156.13\,T^2$ (where L = wavelength in cm, T = period in s).

The height of the wave (H) is controlled by the stroke of the paddle - the greater the distance moved by the paddle, forward through the water, the higher will be the resulting wave.

Wave length and period also control velocity in deep-water waves since $C = \sqrt{gL/2\pi}$, where C = celerity (velocity).

If the depth of water (d) in the wave tank is less than one-half of the wave length, the above equations do not exactly fit the wave parameters. If d/L is between 0.5 and 0.05 the velocity equation will be

$$C = \sqrt{\frac{gL}{2\pi} \tanh \frac{2\pi d}{L}}$$

If $d/L < 0.05$ the velocity equation is $C = \sqrt{gd}$. When wave height (H) becomes greater than $\frac{1}{7}L$, the wave breaks (H/L being the wave steepness ratio where, say, $\frac{1}{150}$ would be *low* and $\frac{1}{15}$ would be *steep*).

The depth of water used in the tank must therefore be taken into account if calculations are to be made, and if breaking waves are required.

Waves may be erosive (destructive) or depositional (constructive) and this characteristic is controlled by the period. In the experiments described by Stong, a tank holding about 15 cm depth of water with a sandy beach sloping at about 15° will produce destructive waves if the period is about 1 s, and constructive waves if it is about 2 s. The waves developed had a height of about 12 cm and were of the steep variety, characteristic of shallow water.

Figure 9.4. A flume-mounted, adjustable-pitch wave generator. (After C. L. Stong.)

Figure 9.5. Variable-period, variable-pitch wave generator, as used for experiments in this chapter (here fitted with narrow paddle).

Experiment 1

Purpose: To produce a landform with destructive waves.

Apparatus: Flume with inspection windows; wave generator; baffle; point gauge; medium sand; timer.

Procedure: A test bed was set up in the flume as shown in Figures 9.6 and 9.11, comprising a beach sloping at about $12\frac{1}{2}°$, 15.0 cm in height and 52 cm in horizontal length, with a deeper-water bed layer. Note that this may most conveniently be constructed from the equilibrium form of a preceding constructive wave experiment such as Experiment 2. The wave generator was set up in a system of clamps with a distance between paddle hinge and beach foot of 1.11 m. Sea-water was run in with care, and the initial surfaces and dispositions were marked with a grease pencil on the side windows (Figure 9.7). The wave generator stroke and period were pre-set.

Controls: Sea-level 6.0 cm; wave period 1 s/wave;

Figure 9.6. Test bed for Experiment 1. Note position of wave generator and the wave-crest point gauge.

Figure 9.7. Experiment 1 before commencement, with beach and water-level positions marked on windows.

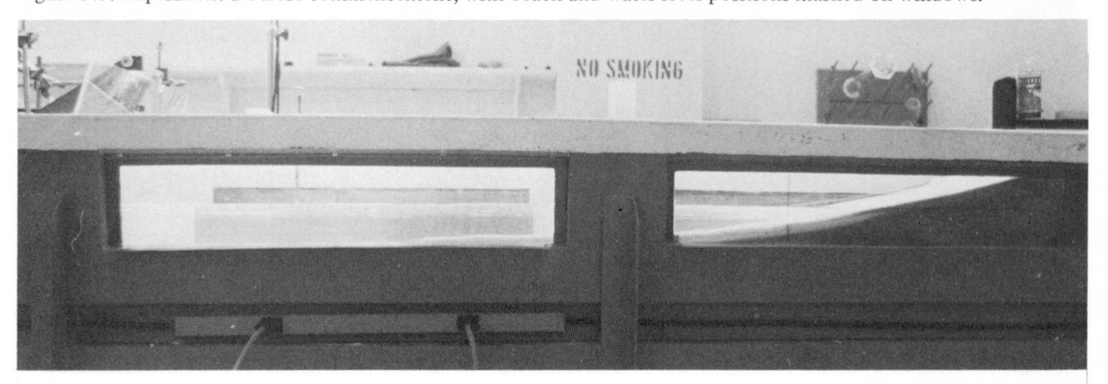

wave amplitude 3.5 cm measured by point gauge; wave celerity about 1 m in 1.2 s by measurement; wave length about 1.0 m by measurement; duration of experiment 63.7 minutes.

Observations and data: The waves broke just after leaving the paddle, and advanced as surf over the sea bed and up the beach. The backwash had a clear run down the beach before meeting the next wave. From commencement the backwash drew material seaward off the beach. By MINUTE 8 sand was being drawn back towards the generator to join a bar of 30 cm transverse length at 60 cm from the foot of the beach. To the immediate landward side of the bar there was an irregular clear patch for 10 cm and then an accumulation of rippled sediment towards the beach. There was a lowering by about 1.0 cm of beach and offshore surfaces below mean sea-level. At MINUTE 15 conditions were generally as at MINUTE 8 (Figure 9.8). At MINUTE 21 conditions generally were still as at MINUTE 8, i.e. equilibrium had been attained. Figure 9.9 was taken at this time, of a wave leaving the paddle. At MINUTE 63.7, switch-off. The general beach form was of considerable erosion with somewhat more on the left than on the right (Figure 9.10 (a) right, (b) left profile). The post-run equilibrium surface was as shown in Figure 9.11, with parts of the initial surface intact up the beach and a micro-cliff at the landward limit of water action. The average beach lowering was 1.0 cm, representing 3410 g of sand.

Conclusions: The effect of waves of 1.0 s period and 3.5 cm amplitude in 6.0 cm deep sea-water, with the other controls given, was to break offshore and to dislodge beach material from the initial surface and to transport it seaward, removing about 1.0 cm depth of sand from 40 cm of beach (zone A, Figure 9.11). This material, with a mass of some 3400 g, was carried by backwash across zone B, where there was some accretion to form a ridge. To seaward of the ridge, a relatively sediment-free trough was formed (zone C) and extensive inshore deposits formed a bar at zone D.

Equilibrium conditions were approached after the first 15 minutes of the experiment, with a gradual slowing down of bedform change, and there was very little alteration in bedform configuration during the last 15 minutes of the run. The equilibrium profile included a micro-cliff at the upper edge of the swash zone, and some sand was carried out to the offshore zone to be deposited in fine ripple markings.

Systems interpretation: The process-response system will be taken to include the inshore and foreshore zone inclusive of the seawater in contact with it (the morphological subsystem). The cascading subsystem is primarily the water mass and energy, but includes some sediment movement. Mass input (water) = total wave mass in the duration of the experiment and may be calculated from wave volume V per unit length of wave per second, where

$$V = H^2 \sqrt{\frac{\pi g}{32L}},$$

$$V = 3.5^2 \sqrt{\frac{3.14 \times 981}{32 \times 100}} = 12\ \text{cm}^2\,\text{s}^{-1}.$$

Figure 9.8. Experiment 1 at minute 15, with offshore bar.

Total volume of water in the waves during the run was $V \times$ (duration in seconds) $\times$ (wave length) $= 12 \times (63.7 \times 60) \times 55 = 2\,522\,520\ \text{cm}^3$, total mass $= 2\,522\,520$ g.

Mass (sand) $= 3410$ g dislodged from the beach and moved back to the inshore and offshore zones, with a small proportion moved out of the system to the rear of the wave generator.

Energy (water): The steepness of the wave is $H/L = 3.5/100 = 0.035$, which is low. The energy equation is therefore $E_t = \frac{1}{8}WLH^2 = (1 \times 100 \times 3.5^2)/(8 \times 10\,197.16)$ J per unit length of wave crest per second $= 0.015\ \text{J cm}^{-1}\ \text{s}^{-1}$, where $E_t =$ kinetic energy per unit of time.

Hence for the duration of the experiment the energy (E_k) was given by

$$\begin{aligned} E_k &= E_t \times (\text{duration}) \\ &\quad \times (\text{distance along wave crest}) \\ &= 0.015 \times (63.7 \times 60) \times 55 \\ &= 3153.15\ \text{J}. \end{aligned}$$

Energy (sand): Movement of sand in a seaward direction is dependent upon wave energy or its transformation, backwash energy. The energy intrinsically attributable to sediment is potential energy of position. An estimate of this can be derived by taking H as $\frac{1}{2}$ initial beach relief $= 15.4/2 = 7.7$ cm. Sand energy (E_p) is therefore $(3410 \times 7.7)/10\,197.16 = 2.57$ J.

When the model was first set up, and immediately after commencement, the components of the morphological subsystem were not in phase with those of the cascading subsystem.* Upon commencement of the run, the system dynamics worked first towards producing in-phase conditions and then, by changes in the morphological subsystem and less obvious changes in the cascading subsystem, it worked towards equilibrium. With the controls used a condition of static

* This is not a case of geomorphic misalliance since the initial landform was wave-produced. If the initial landform had been hand-made a situation of misalliance would have existed at commencement.

Figure 9.9. Experiment 1 at minute 21, showing wave leaving paddle.

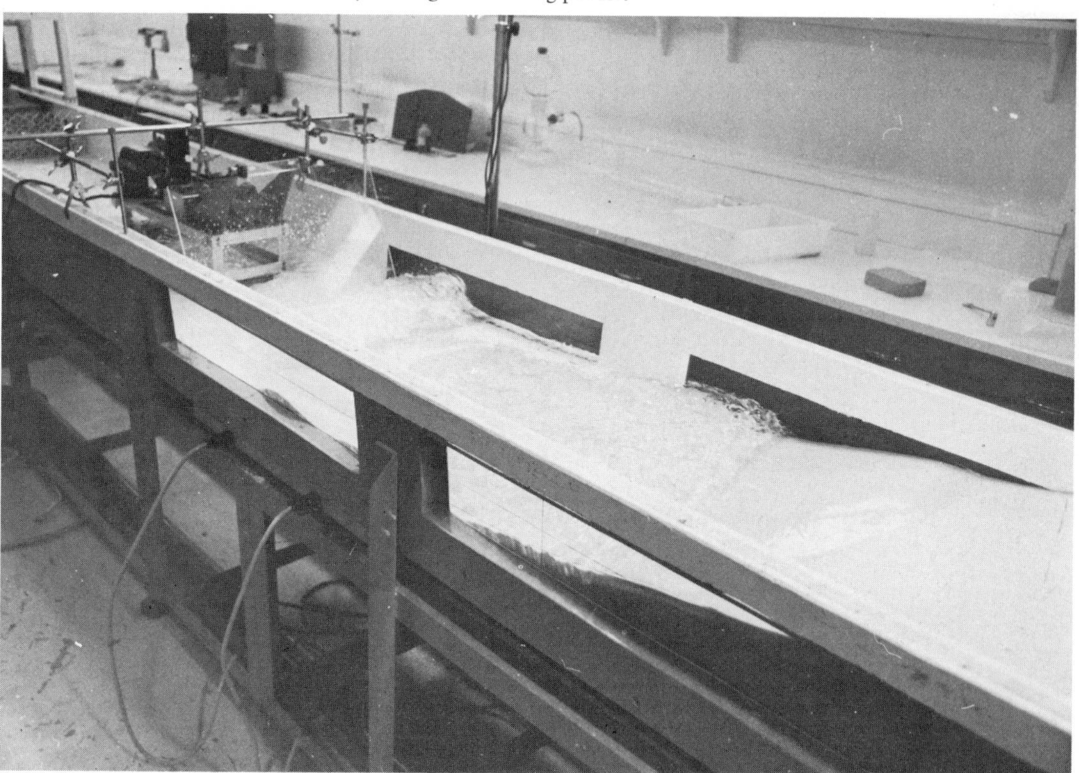

equilibrium was attained within a relaxation time of about 10 minutes. After this stage was reached, movement of sediment within or from the system ceased. Wave characteristics were destructive in the total system environment, which is to say that they caused sediment to move from shore to seaward. A negative-feedback loop existed in which wave energy caused erosion of the beach, and, as erosion increased, backwash transportation increased, shoaling zone slope decreased, causing erosion to decrease. The loop attributes or processes are shown in Figure 9.12. A further link existed between shoaling zone slope and wave characteristics (energy), and backwash transportation of

Figure 9.10 (*a*). Experiment 1 at switch-off (right profile). Compare with Figure 9.10 (*b*).

Figure 9.10 (*b*). Experiment 1 at switch-off (left profile). The left-side of the beach is somewhat more eroded than the right.

Figure 9.11. Mapped beach and sea-bed forms after removal of sea-water, for Experiments 1 and 2.

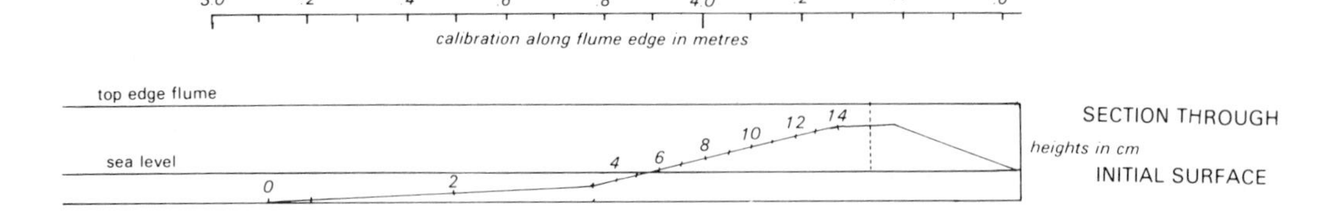

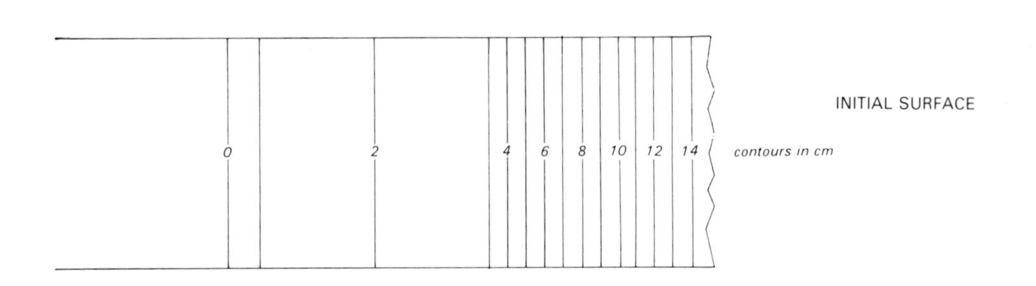

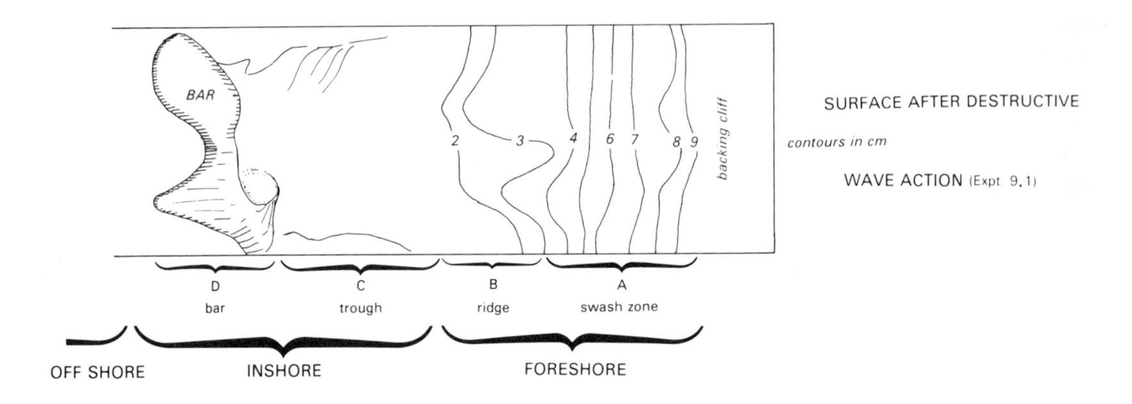

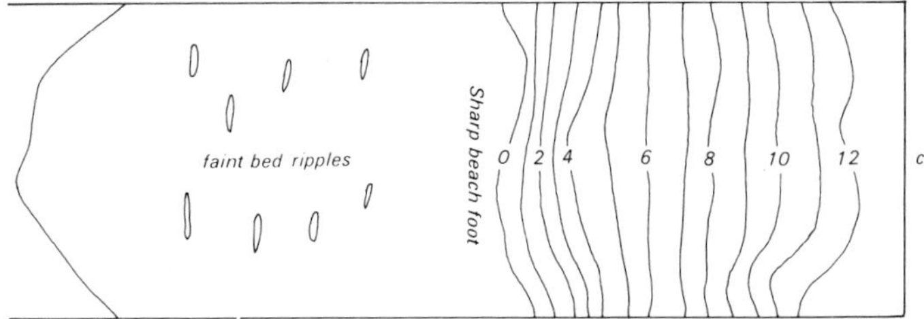

sediment moved some material out of the shore system into the deepwater system.

Experiment 2

Purpose: To produce a landform with constructive waves.

Apparatus: Flume with observation windows; wave generator; baffle; point gauge; medium sand; timer.

Procedure: A landform known to be in equilibrium within a destructive wave system was used – in this case the end form of Experiment 1. The wave generator was left in the same position as for Experiment 1 and sea-water was carefully returned to the flume after completion of measurements for that experiment.

Controls: Sea-level 6.0 cm; wave period 2 s/wave; wave amplitude 3.4 cm before breaking; wave celerity 1 m in 1.2 s by measurement; duration of experiment 23.0 minutes; wave length 2.0 m by measurement.

Observations and data: The wave broke between metre 3.2 (at commencement) and metre 2.6 at MINUTE 15, showing a shift to seaward as the new beach profile developed. The material of the sea bed in the offshore and inshore zones, and the bar, began to migrate landward. Bedforms showed a gentle slope to seaward and a steep slope to landward (Figure 9.13 at MINUTE 19) as opposed to bedforms during destructive wave processes. Material was being driven up the beach by swash, and material from the foot of the beach was lifted leaving a small ledge in that position. By MINUTE 19 a condition of equilibrium was attained with the new beach surface about 1.0 cm above the initial surface and 2.0 cm above the end surface of Experiment 1 (Figure 9.14). The bar and initial trough had disappeared and all sediment on the offshore and inshore zone had been driven onshore, excepting a faint tracery of ripples, as shown in Figure 9.11. The total material driven into the beach was about 7370 cm^3 or 11 423 g.

Conclusions: The effect of waves of 2 s period and 3.5 cm amplitude in 6.0 cm of water, with the other controls obtaining, was to drive material from the offshore, inshore (bar) and foreshore (ridge at beach foot) forward on to the beach, building it up some 2.0 cm higher than at the commencement of the experiment. The swash ran up the beach to a position near the crest of the initial surface, washing out the micro-cliff from Experiment 1 and forming a fairly smooth inclined beach face. Beach profile equilibrium was attained within about 15 minutes of commencement, after which some lateral movement of water began to form secondary features on the beach, producing some modification by switch-off at MINUTE 23. No sea-bed material survived the effect of the long, constructive waves, excepting a faint tracery of bed ripples.

Systems interpretation: The process-response system will be taken to include the inshore and foreshore zone, inclusive of the water involved. The cascading subsystem includes the waves and some sediment movement.

Mass input (water): The equation for volume (mass) is

$$V = H^2 \sqrt{\frac{\pi g}{32L}}$$

per unit length of wave (or cm).

$$V = 3.5^2 \sqrt{\frac{3.14 \times 981}{32 \times 200}}$$

$$= 12.25 \times 0.694$$
$$= 8.5\ cm^2\,s^{-1}.$$

Total volume of water in waves during the run was given by $V \times$ (duration in seconds) $\times$ (distance along wave crest/period $= \frac{1}{2}$ (8.5 x 23 x 60) x 55 = 322 575 cm^3, total mass = 322 575 g.

Figure 9.12. Negative-feedback loop for destructive waves on beach (Experiment 1).

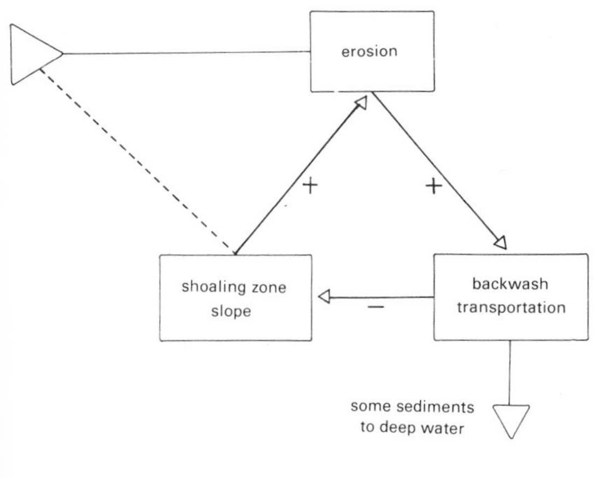

Sand mass in the system dynamics is 11 423 g.

Energy (waves): Steepness of wave is $H/L = 3.5/200 = 0.0175$, which is low. Use

$$E_t = \tfrac{1}{8} WLH^2 = \frac{1 \times 200 \times 12.25}{8 \times 10\,197.16}$$

$$= 0.3 \text{ J cm}^{-1} \text{ per } 2 \text{ s}$$
$$= \tfrac{1}{2}(0.03 \times 23 \times 60 \times 55)$$
$$= 1138.5 \text{ J in 23 minutes.}$$

Energy (sand): Sand is moved upwards onto the beach, and its movement represents work done by wave energy of 11 423 g upwards at an average of 7.7 cm = 87 957 g cm. This represents 87 957/10 197.16 = 8.6 J.

If this value for work done is compared with the total wave energy of the system in 23 minutes (1138.5 J) it is seen that only a small portion of the wave energy was used in re-shaping the beach form, and considerable energy, particularly in the static equilibrium phase, must have been expended in friction and heat.

As in Experiment 1, a negative-feedback loop existed, in which wave energy caused beach accretion, which increased the beach slope, which increased backwash transportation, resulting in decreased beach accretion: a reverse situation to that of Figure 9.12, but similar to the loop described in Chorley and Kennedy's[10] Figure 4.4.

Figure 9.13. Experiment 2 at minute 19 (right profile).

Figure 9.14. Experiment 2 at minute 19 (left profile).

10 The wind tunnel

The action of wind upon land surfaces may be studied in the laboratory by means of a wind tunnel. This aeolian equivalent of the flume has been neglected in instructional situations, probably because of the size of equipment described in the literature. Bagnold,[1] whose work on the physics of blown sand remains the primary reference, used a 30 ft tunnel, and other workers such as Kawamura have used much longer apparatus, as described by King[2] in her review of experimental techniques in wind process studies.

A wind tunnel designed for the physical geography laboratory of Memorial University, uses Bagnold's essential principles, but incorporates the technique of re-circulation found useful in flume construction. The need was seen to fit about 9 m (30 ft) of tunnel into the available laboratory space; to reduce noise levels of open-ended air-intakes or exits, and to eliminate air movement within the laboratory in the vicinity of the wind tunnel. The apparatus developed rests on a 4.57 m × 1.52 m (15 ft × 5 ft) table, built to laboratory-bench height. It is almost silent, except at highest wind speeds, and produces little external air movement even with an open sand trap.

The design detail is given in Figure 10.1, with general proportions and materials shown in Figures 10.2 and 10.3. The table was constructed of dimensional lumber with a plywood top, made rigid with triangular-section leg-webbing. The tunnel was made of plexiglass for maximum visibility, but could have been of plywood with inspection panels. The fan housing was made of plywood for rigidity and strength (Figure 10.4). A sand trap was built in just behind the fan housing, by cutting out a large floor section, fitting in guiding vanes to edge sand into the trap, and then constructing a well to ensure that sand dropped into a large tray on the floor beneath.

The fan was a $\frac{1}{2}$ h.p. variable-speed extractor model with a nominal air transfer rate of 174 m^3/min (6200 ft^3/min), but with an effective test-section rate of 122 m^3/min (4300 ft^3/min), giving through the 40 cm × 40 cm test section maximum wind speeds of 45 km/hr (28 miles/hr).

Wind speeds were checked, for calibration purposes and during the pre-set stage of an experiment, with a hand anemometer (Figure 10.5), and a manometer was constructed using the tank and inclined tube system described by Bagnold (p. 41 of reference 1). The velocity and static tubes were fitted into the curved segment of the tunnel just upwind from the test bed, and coloured alcohol was used as the liquid (Figure 10.6). The velocity tube was constructed of 1 mm internal diameter glass capillary, cement-bonded into a wider soft-glass tube. The manometer was calibrated while the anemometer was placed in the test bed, and thus gave test bed values.

The test area was 39.5 cm × 39.5 cm square, and 2.44 m (8.0 ft) long with a lift-off lid. Owing to a lack of pressure differential there was no tendency for the lid to rise. The floor of the test section was lined with a mixture of basalt, ground to medium sand size, and carborundum sand. This enabled sand to anchor and provided a good background for observation and photography.

The sand used in the models was very clean, dry silica sand sieved out to between 0.5 and 0.177 mm. It was worked with scrapers and an adjustable former blade (Figure 10.7) to which a card was attached to tidy up the edges. Brushing and gentle blowing smoothed over sharp edges and small pits.

Leading air into the test section from a bend created a velocity differential and biased eddying at the head of the test section, but the effect was reduced by placing baffles into the tunnel at the end of the bend. Two adjustable baffles were found to be adequate to deflect air into a more uniform path (Figure 10.5), and, as may be seen from the rippled surface of Experiment 1, little distortion then occurred.

The experiments described below employ a variety of wind speeds. In each case the whole apparatus must be cleaned out before setting up, and sand may be removed from the sand trap and awkward corners with a vacuum cleaner.

The sand hopper, when used, fed into the wind tunnel by gravity through a pipe which rested on the tunnel floor near the manometer velocity tube. The pipe had a V-notch cut into its lower edge, and wind action was sufficient to keep the end clear and allow sand to trickle out. It is doubtful whether a sand feed of such low delivery significantly affects the dynamics of the experiments described, but it demonstrates the ability of systems to incorporate an external supply of sand.

The experiments included demonstrate the action of wind upon sand surfaces, showing that at lower velocities ripples form, and that obstructions on a plane sand surface produce distortion of the ripple pattern - and, it may be hypothesized, play a role in the development of less common arid landforms. Analogue forms are not suggested in this chapter and a scale of 1:1 may be assumed, including a micro-barchan. With due attention to process, it is possible to scale features up and refer to large desert dunes and deflation hollows.

There is little literature appropriate to the use of the wind tunnel in landform studies. Bagnold[1]

Figure 10.1. Design detail for wind tunnel, as used for experiments in this chapter.

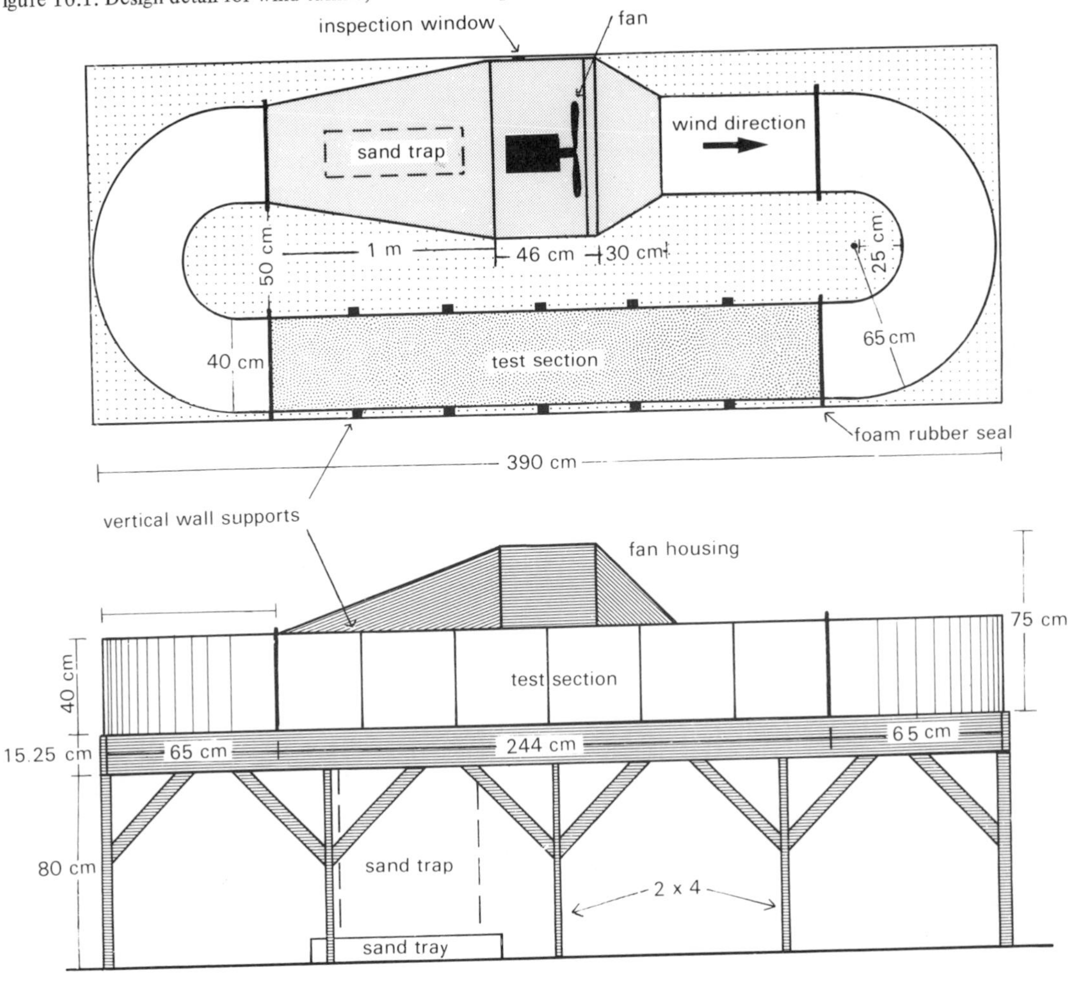

Figure 10.2. Wind tunnel from up-wind end.

Figure 10.3. Wind tunnel, showing test section with black sand test floor.

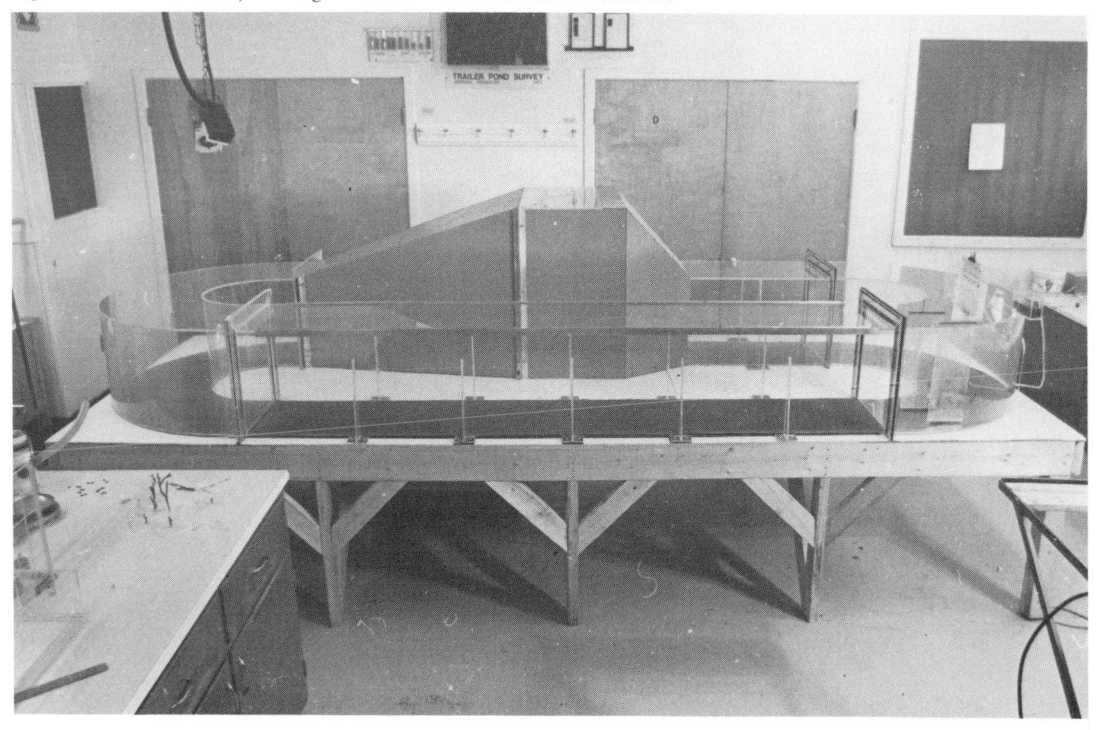

and Cooke and Warren[3] should be consulted for process discussion and Ritter[4] offers a theoretical summary and nomenclature at the undergraduate level.

Experiment 1

Purpose: To produce a wind-formed surface.
Apparatus: Wind tunnel; dry medium 'A' sand; hopper; carriage and former blade; timer.
Procedure: The wind speed was pre-set with a hand anemometer on the test bed; clean, sieved sand was weighed into two parts - 13.6 kg into the test bed, and 1.544 kg into the hopper. The sand bed was made up to 1.2 cm thick, 39.5 cm wide from wall to wall and 140.8 cm long. The edges against the plexiglass walls were trimmed with a card taped to the former blade (Figure 10.7). The windward and lee edges of the sand bed were squared off and trimmed (Figure 10.8 (*a*) and (*b*)). The lid was placed on the test section; the sand edges were marked with a grease pencil on the plexiglass wall; windspeed was monitored throughout the experiment by a manometer.
Controls: Sand grade 0.5-0.177 mm; wind speed 13 km/hr; duration 33.8 minutes.
Observations and data: The windward edge began to erode immediately after commencement, with some accretion from the sand hopper; sand moved over the bed by traction, saltation and suspension. Very fine grains remained in suspension with a general height limit of 6.5 cm above the sand surface, but occasionally rising to as high as 8.7 cm. Saltation trajectories were at a maximum of 6.5 cm in height but were mostly less, and traction movement was in short starts. Ripples formed within a minute of switch-on and rapidly assumed a wavelength of 5.0 cm, and were 2 or 3 mm deep, transverse to the wind direction, with a gentle upwind surface and steep slip face, giving a height/length ratio of 1:20 (Figure 10.9).

The windward edge of the sand bed advanced at a rate of 0.51 cm/min, while the lee edge advanced at 0.22 cm/min, producing an aeolian surface of ever-decreasing length. Ripples advanced over the bed surface at 0.33 cm/min. During the 33.8 minute run the upwind portion of the sand bed surface was lowered by up to 0.6 cm, while the downward portion gained up to 0.15 cm in

Figure 10.4. Fan housing on wind tunnel.

Figure 10.5. Hand anemometer, in position in test section.

thickness (Figure 10.10, (*a*) and (*b*)). At the end of the experiment the test bed contained 9.407 kg of sand; the hopper, 0.808 kg; and the sand trap and surrounds, 4.929 kg.

Conclusion: With a wind speed of 13 km/hr and sand grains of 0.5–0.177 mm, the drag velocity is well above the fluid threshold of about 4.0 km/hr, as determined by

$$V_t = A \sqrt{\frac{\rho_s - \rho_a}{\rho_a} \times gD},$$

where V_t = fluid threshold (velocity of wind required to initiate movement of sand grains in use); A is a constant, of value 0.1 for air; ρ_s is the density of sand $= 2.67\ \text{g cm}^{-3}$; ρ_a is the density of air $= 0.00129\ \text{g cm}^{-3}$ at normal pressure and 0°C; g is the acceleration due to gravity; D is the mean diameter of the sand grains, taken as 0.025 cm. Hence

$$V_t = 0.1 \sqrt{\frac{2.67 - 0.00129}{0.00129} \times 981 \times 0.025}$$

$$= 112\ \text{cm s}^{-1}$$

$$= \frac{112 \times 60 \times 60}{100\,000} = 4.0\ \text{km/hr}.$$

Figure 10.6. Manometer, in position in upwind section of wind tunnel.

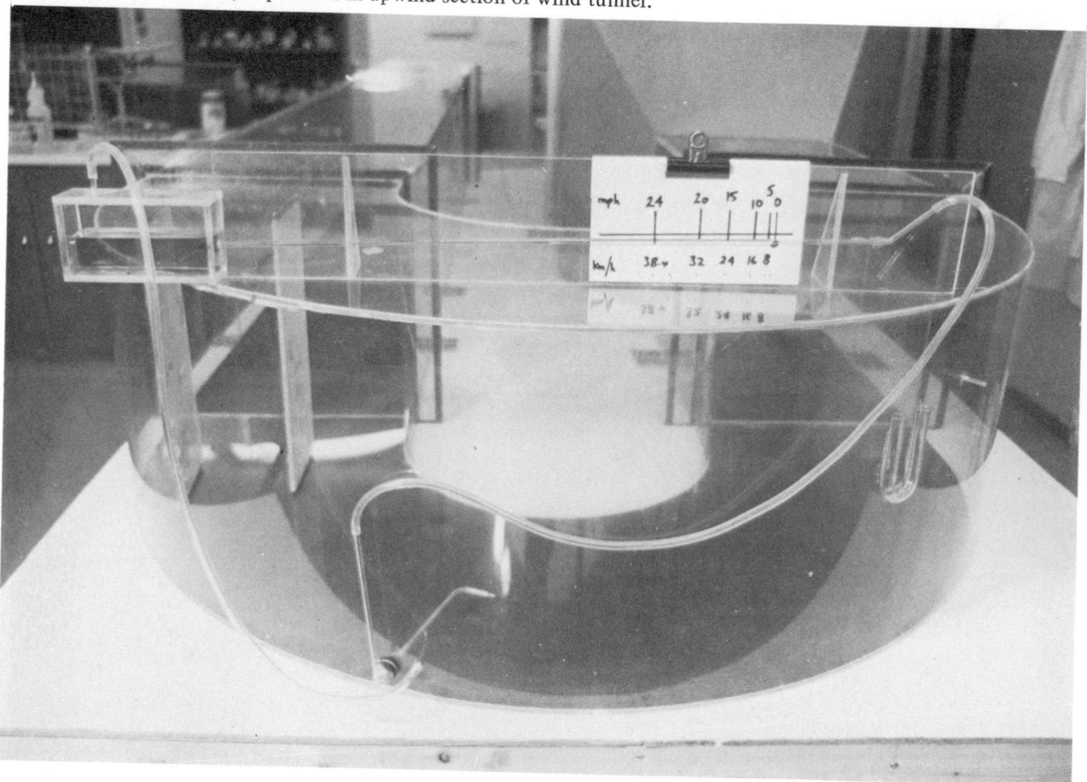

Figure 10.7. Sand former.

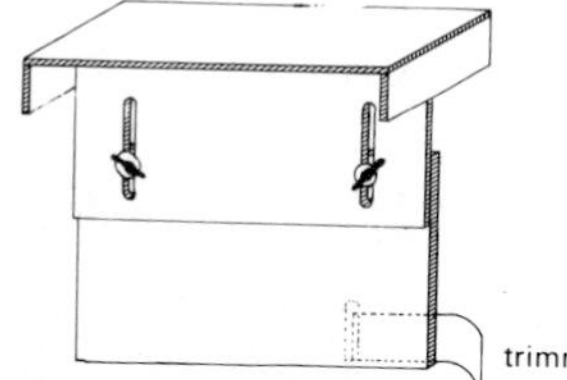

Figure 10.8 (*a*). Test bed for Experiment 1 (side view).

Figure 10.8 (*b*). Test bed for Experiment 1 (plan view and upwind to right).

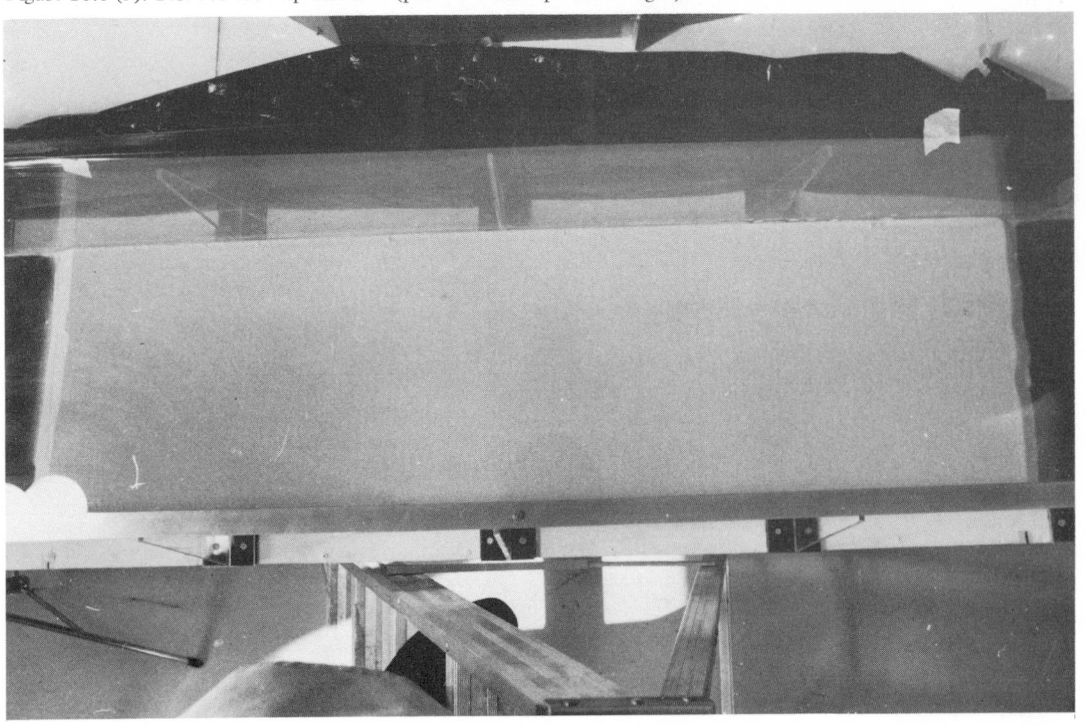

Sand was transported immediately upon commencement of the run. The very finest particles were lifted up to 6.5 cm and carried in suspension, but saltation was the commonest mode, and the largest particles moved by rolling or sliding. The windward edge of the sand bed retreated at about twice the rate of the lee edge, while ripples advanced at an intermediate rate. Ripples assumed a fairly regular wave length and the H/L ratio of 1:20 approximated theoretical values. Slip faces were always steeper than upwind faces and orientation was transverse to wind direction. It may be concluded that, with sand size and other controls given, low wind speeds will produce a surface of uniform ripples transverse to wind direction.

Systems interpretation: The total bed will be regarded as a system together with the wind layer up to 10.0 cm above the sand bed surface in which the wind load was contained. The initial morphological subsystem was, therefore, 140.8 cm long, 1.2 cm deep and 39.5 cm wide. Its mass was 13.6 kg. The cascading subsystem was an air mass of 395 cm^2 section moving over the bed at 13 km/hr, representing a total volume given by

volume

$$= (\text{cross-section}) \times (\text{wind speed}) \times (\text{duration})$$
$$= 395 \times (13 \times 10^5) \times (33.8/60)$$
$$= 2.89 \times 10^8 \text{ cm}^3 \text{ in 33.8 minutes.}$$

Mass

$$= \text{volume} \times \text{air density at } 20\,°\text{C}$$
$$= 2.89 \times 10^8 \times 0.00125$$
$$= 3.616 \times 10^5 \text{ g in 33.8 minutes.}$$

Energy of the air mass is given by

E_k

$$= \tfrac{1}{2}(\text{mass}) \times (\text{velocity})^2$$
$$= \tfrac{1}{2} \times (3.616 \times 10^5) \times 361^2$$
$$= 2.36 \times 10^{10} \text{ ergs}$$
$$= 2.36 \times 10^3 \text{ J.}$$

The mass of sand in the cascading system was the difference in bed mass between commencement (13.6 kg) and end (9.407 kg), equal to 4.193 kg. Note that the system gained 0.736 kg from the hopper, without which bed loss would have been 4.929 kg, and sand moving on the sand bed surface, but not off it, is also an unmeasured part of the cascading subsystem. Sand energy is derived from wind energy and, strictly speaking, should be deducted from wind energy. However, the modes and rates of sand movement are unknown and would require considerable experimental analysis to determine, so that it is convenient to recognize simply that 2.36×10^3 J is total energy inclusive of sand energy.

As the test bed was transformed in shape it offered a progressively smaller, rough, trapping

Figure 10.9. Experiment 1 soon after commencement (upwind to right).

Figure 10.10 (*a*). Experiment 1 at end of run (side view). Note ripples showing strongly against plexiglass to right of centre.

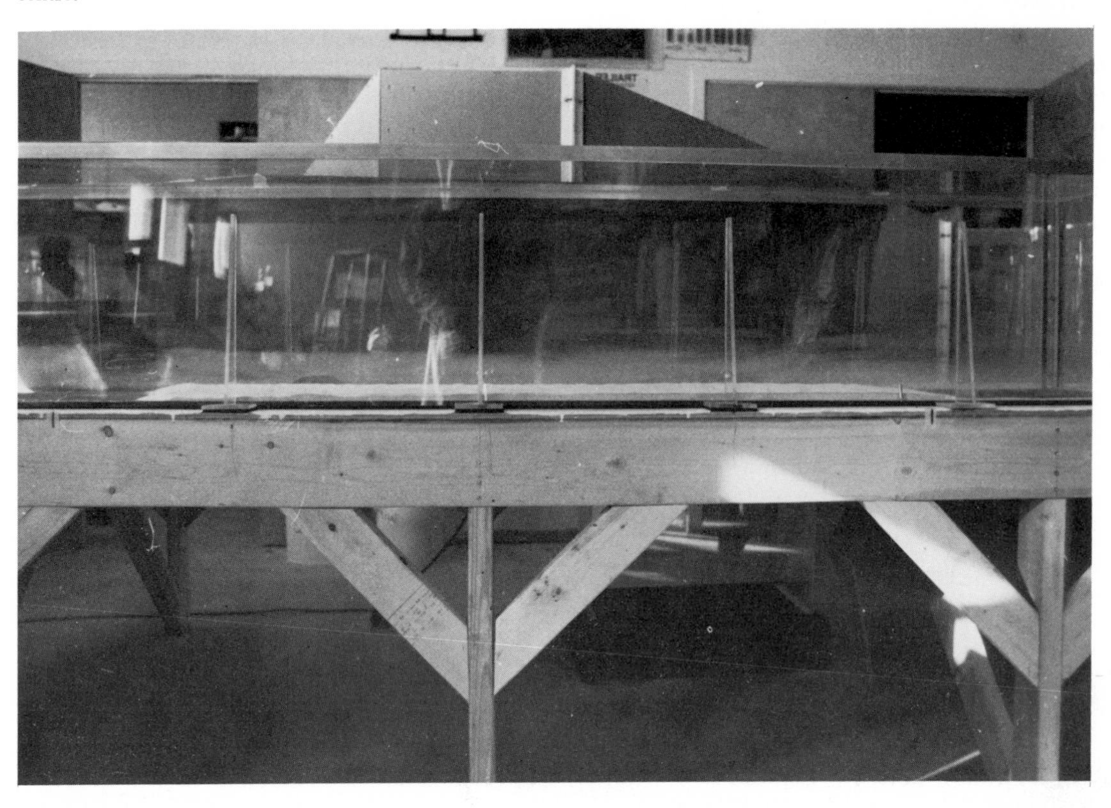

Figure 10.10 (*b*). Experiment 1 at end of run (plan view).

surface to dislodged particles, which shot out of the system once beyond the lee edge. In this finite model, therefore, a positive-feedback loop existed (Figure 10.11) in which the greater the development of wind-formed surface, the more rapid the erosion. As such a surface continued to develop, deflation hollows and highly irregular windward and lee edges occurred and the increasing rate of erosion was readily observed.

Figure 10.11. Positive-feedback loop (Experiment 1).

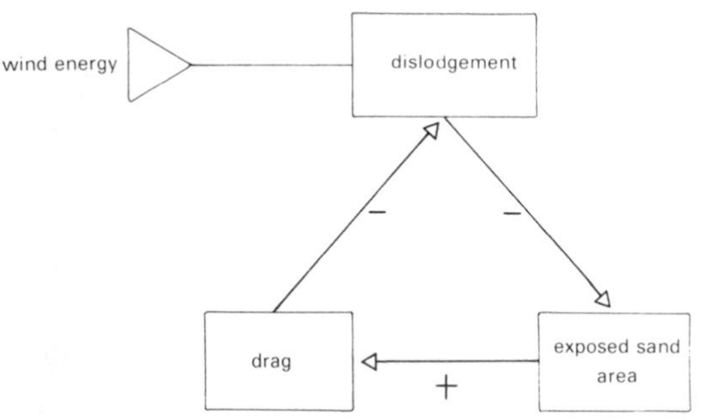

Experiment 2

Purpose: To produce obstruction effects on a wind-formed surface.

Apparatus: Wind tunnel; medium 'A' sand; former blade; eight twigs cemented to 5 cm x 5 cm x 2 mm thick bases; timer, etc.

Procedure: Wind speed was pre-set. A sand bed was made up with 13.951 kg of sand, 1.43 m long, 39.5 cm wide and 1.2 cm deep, smoothed and trimmed with the former, and obstructions were slid in beneath the sand to make firm contact with the test bed floor. Sand was carefully replaced over the bases of the obstructions, and smoothed with a card and by gently blowing through a tube. 1.544 kg of sand was placed in the hopper. Sand levels were marked on the tunnel walls and notes were made of obstruction positions (Figure 10.12).

Controls: Wind speed 16 km/hr; obstacles were set at least 7.0 cm from the walls and 12 cm from each other (Figure 10.13); duration of run 33.8 minutes; sand grade 0.5–0.177 mm.

Figure 10.12. Test bed for Experiment 2, with obstructions in downwind sector.

Figure 10.13. Location of twigs (with heights in cm) in test bed (Experiment 2).

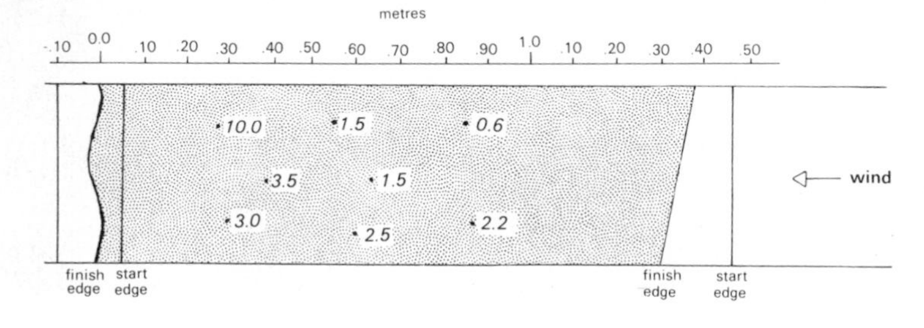

Table 10.1. *Movement of sand bed edges*

	Movement of edge	
Time (min)	Upwind edge (cm)	Lee edge (cm)
7	5.0	0.8
16	8.0	1.5
27	12.5	4.0
33.8	14.5	5.5

Observation and data: Ripples developed within the first minute of the run. For some minutes the obstructions made no impression on the ripple surface and by MINUTE 20 (Figure 10.14) the ripples were still transverse with little distortion. The upwind and lee edges of the sand bed advanced throughout the run as shown in Table 10.1. There was an average movement of 0.42 cm/min for the upwind edge and 0.16 cm/min for the lee edge. The wavelength of the ripples was between 3 and 5 cm averaging 4.5 cm, and the hollows were 0.2 and 0.3 cm deep, giving an H/L ratio of 1:18 (Figure 10.15). The ripples advanced at a rate of 0.53 cm/min.

After MINUTE 20 the effect of obstructions was becoming more obvious. Where ripple crest-lines advanced upon obstacles the lines became distorted (Figure 10.16 (*a*)). Between obstacles distorted wave trains formed latticed interference patterns (Figure 10.16 (*b*)). Pitting occurred at the base of the obstructions (Figure 10.16 (*c*)), and a sand rib developed in the wind shadow (Figure 10.16 (*d*)). These features are visible in Figure 10.17. At the conclusion of the experiment, the sand trap and surrounds contained 4.07 kg of sand, the hopper contained 0.86 kg, and the sand bed 10.1 kg.

Conclusions: With the controls used, a ripple-covered aeolian surface was produced similar to that of Experiment 1, with wavelength 4.5 cm and depth 2.5 mm, transverse to wind direction and advancing at a rate of 0.53 cm/min. The ripple pattern tended towards an equilibrium

Figure 10.14. Experiment 2 at minute 20.

condition, but, with time, became progressively more distorted as the obstructions interfered with the basic aerodynamics. Ripples took a diagonal orientation and crossed each other; pits developed in strong eddy zones around the bases of obstructions, and sand ribs formed in the wind shadow of each obstruction. The sand bed lost material from the upwind edge and gained on the lee edge, with a net loss of material. The effect of the obstructions was seen to be local and superficial during the 33.8 minutes of the run, affecting micro-features rather than the gross sand bed.
Systems interpretation: The morphological subsystem is essentially the same as in Experiment 1,

Figure 10.15. Upwind sector at about minute 2, showing ripple form against side of viewing panel.

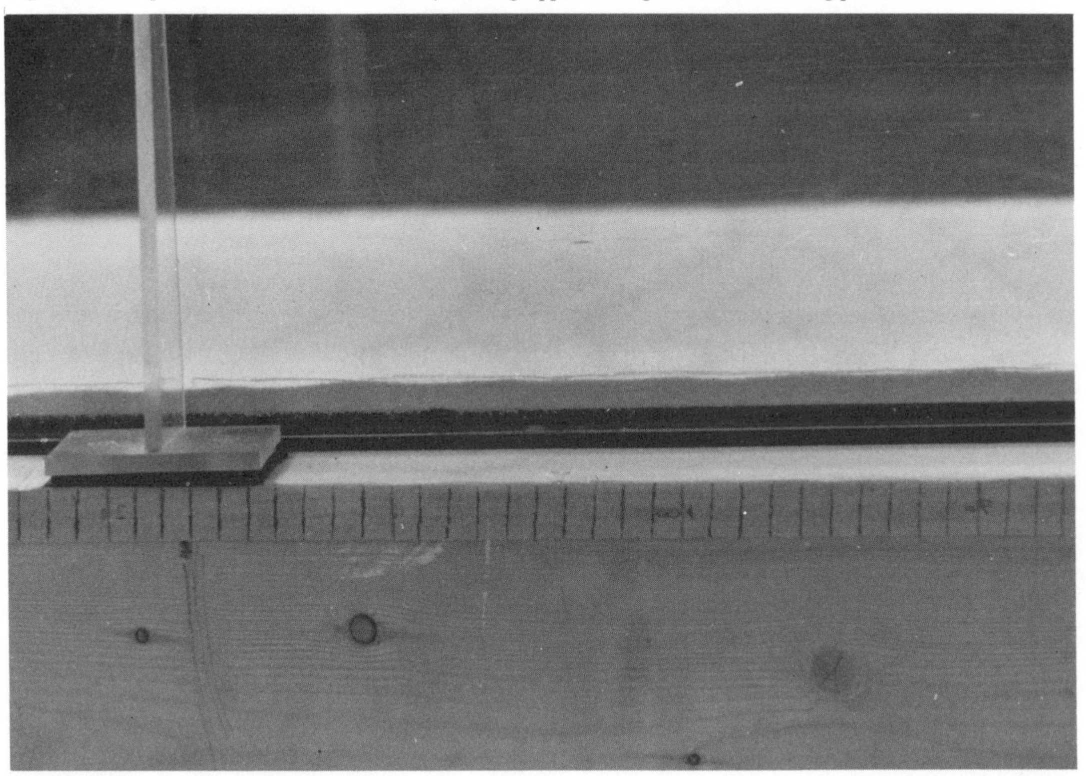

Figure 10.16. The relationship of ripple form to obstructions, showing pitting and sand ribs.

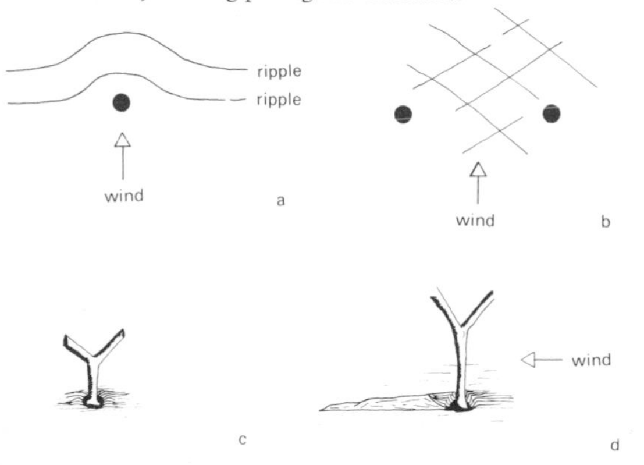

excepting the addition of obstructions and the eventual distortion of the ripple subsystem. The cascading subsystem, taking a 10 cm deep air mass above the sand surface is as follows.

Volume (air)

$$= (\text{cross-section}) \times (\text{wind speed}) \times (\text{duration})$$
$$= 395 \times (16 \times 10^5) \times (33.8/60)$$
$$= 3.56 \times 10^8\,\text{cm}^3 \text{ in } 33.8 \text{ minutes.}$$

Mass (air)

$$= \text{volume} \times \text{air density at } 20^\circ\text{C}$$
$$= 3.56 \times 10^8\,\text{cm}^3 \times 0.00125$$
$$= 4.45 \times 10^5\,\text{g in } 33.8 \text{ minutes.}$$

Mass (sand) in sand trap and downwind bend $= 4.07 + 0.46 = 4.53$ kg. Energy (wind) is given by

E_k

$$= \tfrac{1}{2}(\text{mass}) \times (\text{velocity})^2$$
$$= \tfrac{1}{2} \times (4.45 \times 10^5) \times \left(\frac{16 \times 10^5}{60 \times 60}\right)^2$$
$$= \tfrac{1}{2} \times 4.45 \times 10^5 \times 1.925 \times 10^5$$
$$= 4.394 \times 10^{10}\,\text{ergs}$$
$$= 4.394 \times 10^3\,\text{J}.$$

Energy of sand (x J) is derived from wind energy and, for convenience, may be regarded as subsumed by wind energy. The system may be represented as in Figure 10.18.

Experiment 3

Purpose: To produce a barchan-like landform.
Apparatus: Wind tunnel; hand anemometer; medium 'A' sand; funnel; timer. balance.

Figure 10.17. Experiment 2 at end of run.

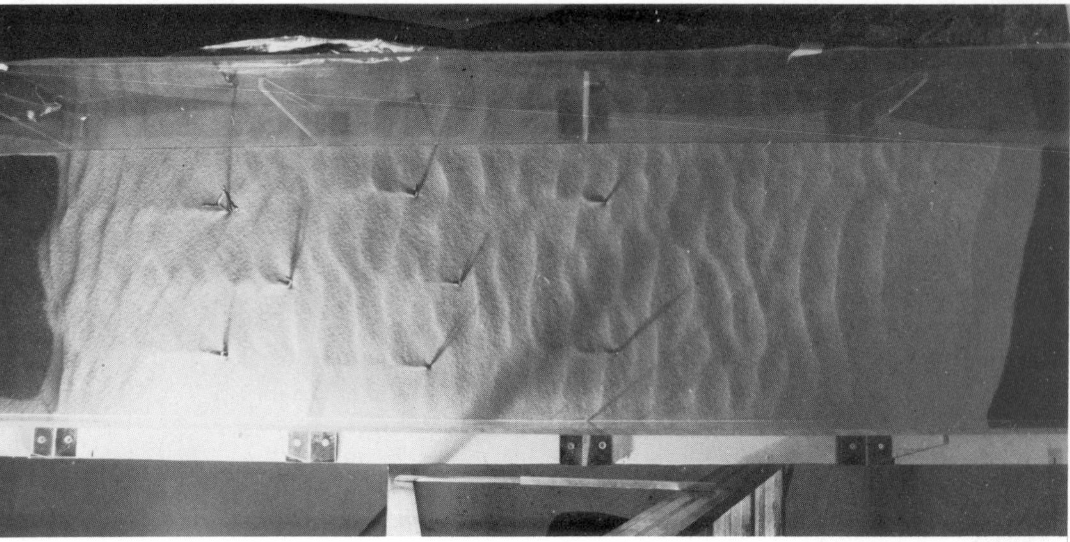

Figure 10.18. Canonical diagram for system in Experiment 2: sand bed and obstructions tend constantly to diminish at upper surface and edges.

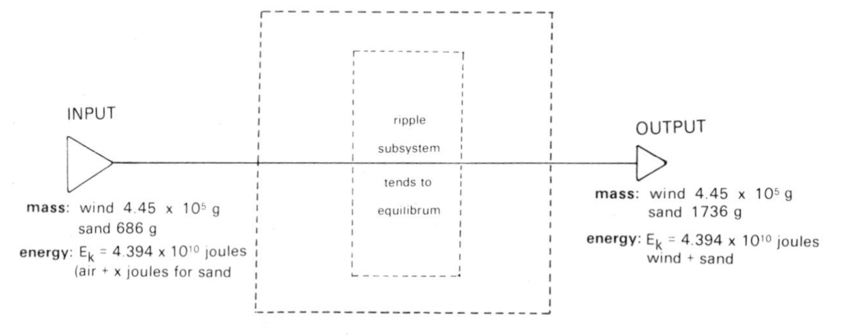

Procedure: Wind speed was pre-set by hand anemometer and manometer; 2000 g of dry, clean medium sand were placed on the test bed by pouring through a funnel to form a cone 25 cm in diameter, and 8.3 cm high (Figure 10.19 (*a*), (*b*) and (*c*)). During the run the experiment was stopped at intermediate times – MINUTES 0.54, 1.31 and 1.536 – for measurement and photography.

Controls: Wind speed 38.6 km/hr; duration 1.783 minutes.

Observations and data: From the moment of commencement, dislodgement of sand from the cone was strong. At MINUTE 0.54 an early stage

Figure 10.19 (*a*). Test bed for Experiment 3 (plan view).

Figure 10.19 (*b*). Test bed for Experiment 3 (side view).

of dune-shaping was seen, with ridge and incipient slip face. Rudimentary horns were appearing (Figure 10.20). By MINUTE 1.31 further development of the MINUTE 0.54 form had occurred with elongation - more pointed into the wind, and slightly more pronounced horns (Figure 10.21). By MINUTE 1.536 the horns were more pronounced, a deep lee cavitation had appeared, and the slip face was broadly concave (Figure 10.22). By MINUTE 1.783 (end point) the barchan form was well developed, with pronounced, blunt horns, a heart-shaped general dune form, a concave slip face and a rounded spine (Figure 10.23 (*a*), (*b*) and (*c*)).

The micro-dune changed form rapidly during the experiment in both size and shape, and also changed position, moving downwind as shown in Table 10.2. The data (Figure 10.24) show an apparently constant rate of movement for all parts of the barchan from MINUTE 1.31 to the end of the run, but with a slightly more rapid rate for the upwind edge than for the horns.

Table 10.2. *Movement of barchan*

	Distances from initial position			
MINUTE	Upwind edge (cm)	Crest (cm)	Lee edge (cm)	Horn (cm)
0.00 (cone)	0.0	12.5	25.0	–
0.54	–	16.2	–	27.0
1.31	3.2	18.1	–	29.3
1.536	7.0	20.6	–	31.3
1.783 (barchan)	9.2	22.4	27.6	32.4

Conclusions: At a wind speed of 38.6 km/hr, air impinging upon a cone of sand causes rapid deflation on the upwind face and incipient accretion on the lee face. The cone loses its initial symmetry within a fraction of a minute and simultaneously migrates downwind. Material is lifted up the backslope by a traction process which is the resultant of forward motion and gravity, the forward motion

Figure 10.19 (*c*). Experiment 3 before commencement.

Figure 10.20. Experiment 3 at minute 0.54, showing early stage of dune formation.

Figure 10.21. Experiment 3 at minute 1.31: dune elongating and horns beginning to develop.

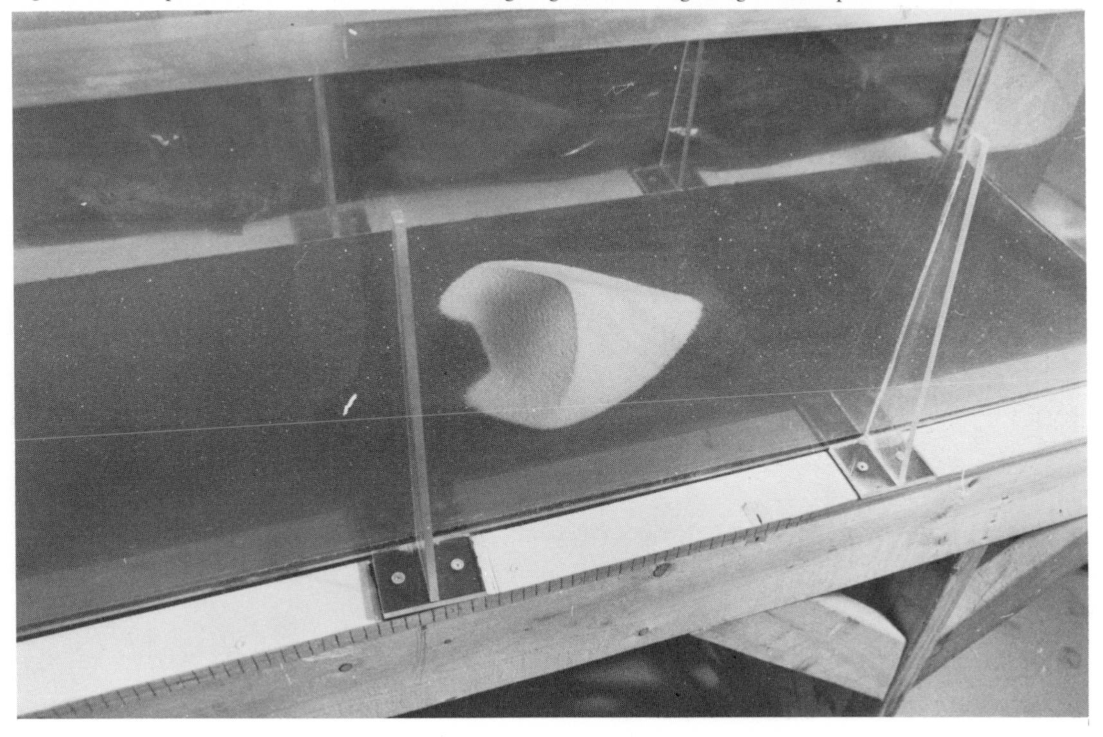

Figure 10.22. Experiment 3 at minute 1.536: lee cavitation developing.

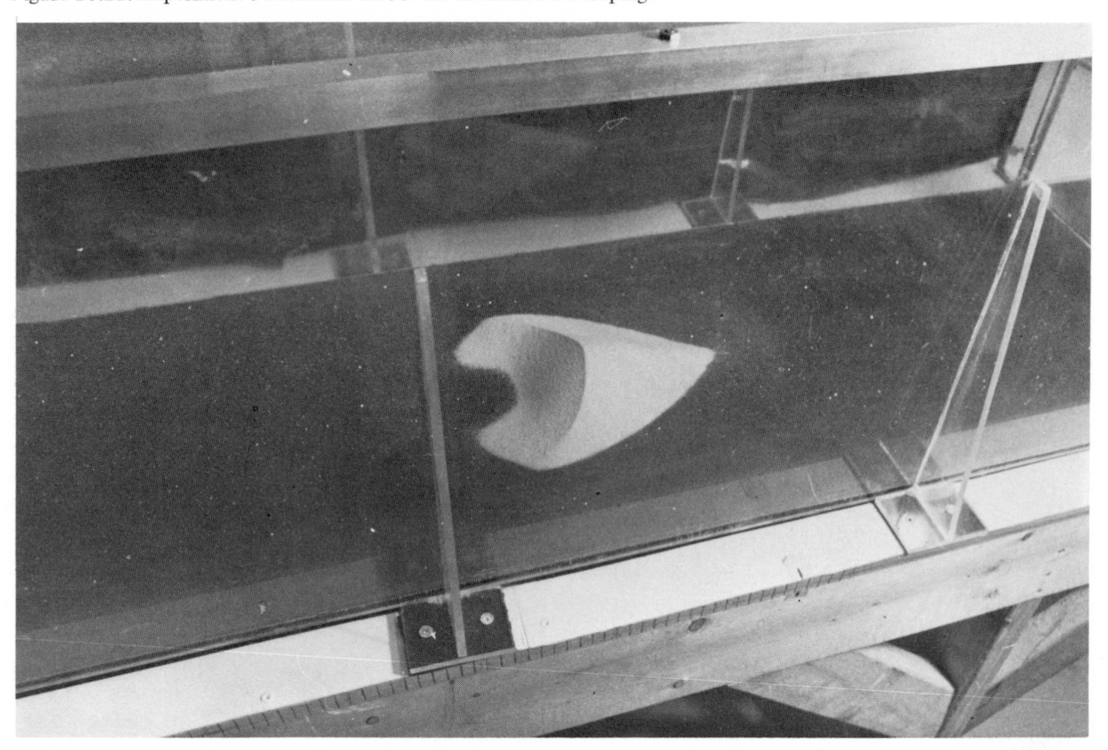

Figure 10.23 (*a*). Experiment 3 at end point (side view), showing wasted mass, with low elevation and concavity in lee.

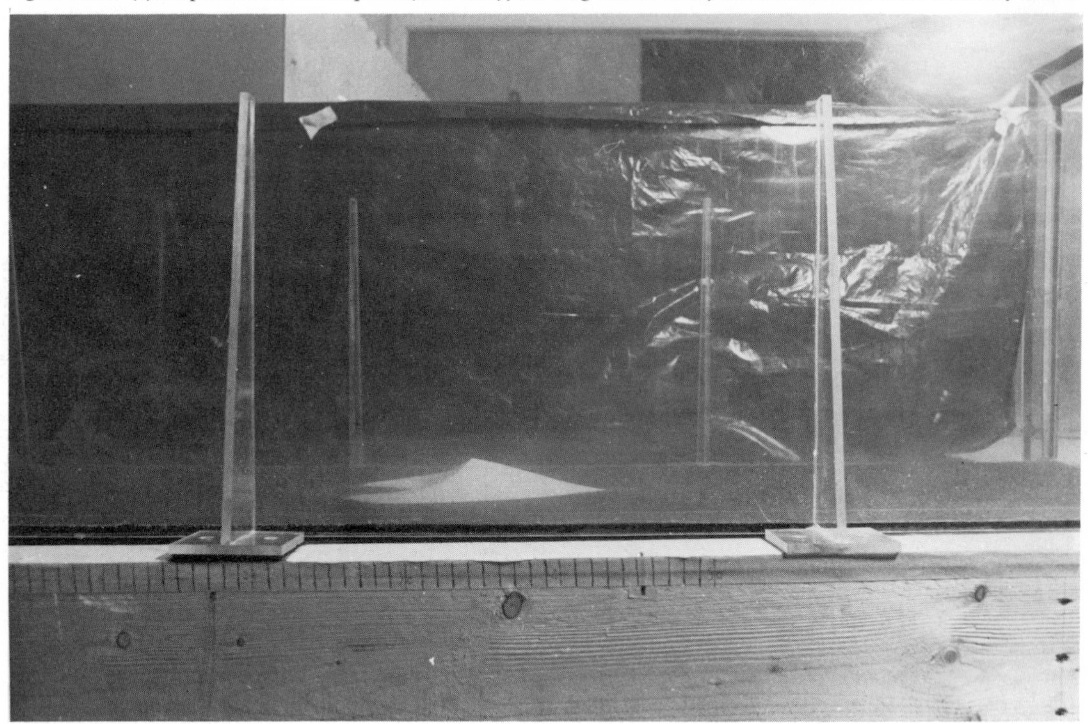

Figure 10.23 (*b*). Experiment 3 at end point (oblique front view).

tending to lift the particle upslope, while gravity counteracts this and stream-lines carry the particle round and down on a gentle trajectory. The result is the development of a sharp transverse ridge, separating the backslope from the lee slip face, down which the material falls, and a less distinct longitudinal ridge as air currents separate to either side of the mass.

On the lee side of the dune strong eddying occurs, which sweeps out material from the base but allows it to collect to either side, with new material falling in from the transverse ridge above. The result is the characteristic barchan concave lee face with horns to either side. The plan of the dune is heart-shaped with a sharp point upwind. Erosion of the dune is rapid, despite some re-circulating airborne sand (or perhaps because of . . .), and at the end of 1.783 minutes the dune mass is only 308.1 g, representing an average attrition rate of 15.8 g s^{-1}.

Systems interpretation: The system changes form rapidly, and crosses a form threshold between cone and barchan.

Mass (sand): originally 2000 g, but 308.1 g at termination. With high wind velocity, finer particles of sand are airborne and re-circulate.

Figure 10.23 (*c*). Experiment 3 at end point (plan view), showing perfect axial symmetry.

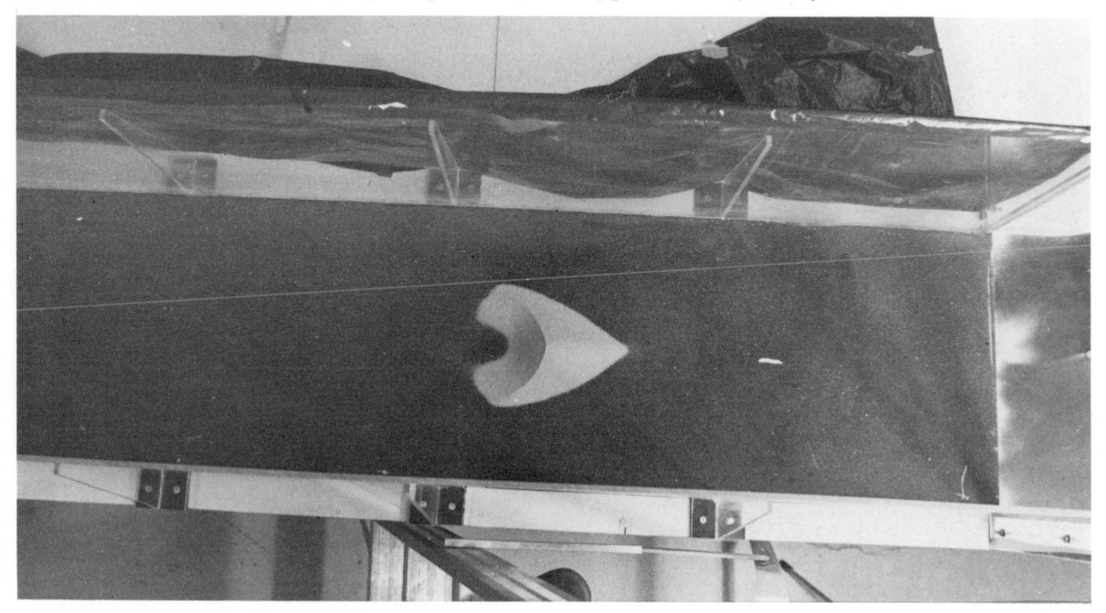

Mass (air): the operative section is taken to be the whole 39.5 cm × 39.5 cm section, since sand is involved at all levels, giving an area of 1560.25 cm^2.

Wind velocity is 38.6 km/hr = 38.6×10^5 cm/hr or $(38.6 \times 10^5)/60$ cm/min = 64 333 cm/min.

The volume of air passing the cone in 1.783 minutes is, therefore, 1560.25 × 64 333 × 1.783 = 1.79×10^8 cm^3, and the mass is $1.79 \times 10^8 \times 0.00125$ = 223 712 g.

Energy (wind) is given by

$$
\begin{aligned}
E_k &= \tfrac{1}{2}(\text{mass}) \times (\text{velocity})^2 \\
&= \tfrac{1}{2} \times (223\,712) \times (1072)^2 \\
&= 1.285 \times 10^{11} \text{ ergs} \\
&= 1.285 \times 10^4 \text{ J.}
\end{aligned}
$$

Sand energy is essentially transposed wind energy, excepting the case of sand particles being struck and driven directly by the fan blades. There is no assessment of the quantity of airborne sand nor of the effect of direct fan impact. Sand energy is thus included as x J in wind energy. Over 12 000 J in less than 2 minutes is an exceptionally high energy system amongst the laboratory models discussed in this work. The dune is formed within about half a minute of commencement and then retains its characteristic shape, while changing slightly in form detail and eroding rapidity. In terms of form therefore, it is possible to recognise a near-equilibrium condition after half a minute, but the system mass decreases. This state may be a manifestation of dynamic equilibrium where fluctuations about a mean shape are not readily discernible, but the trajectory of the *average* states is obvious, as may be seen on the photographs.

Figure 10.24. Advancing features of dune at points in time (Experiment 3).

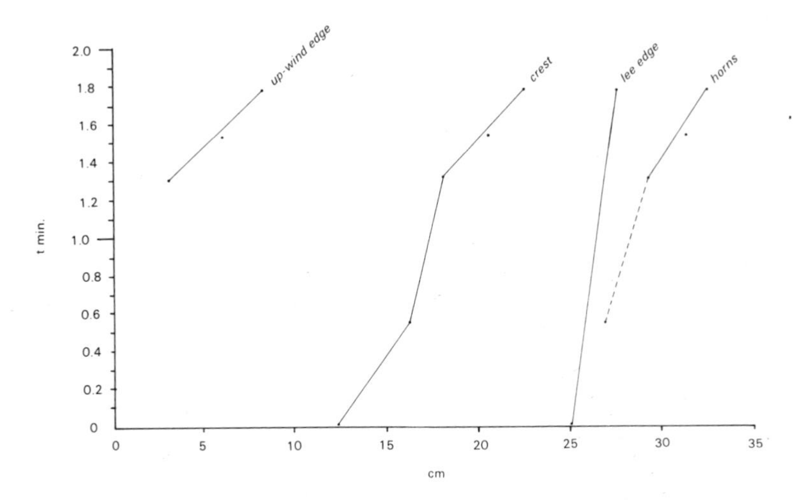

11 Plastic deformation models

The landforms and processes dealt with in other chapters are exogenetic, depending upon forces and materials at the Earth's surface. Within the geological sector of earth science considerable attention has been paid by many workers to the subsurface or endogenetic processes that produce crustal deformation and are responsible in the final analysis for what are taken to be initial landforms.

The precise sequence of events in crustal movements is imperfectly known, but convection currents in the mantle, loading by sedimentation in geosynclines, or unloading during the melting of an ice cover are frequent causes, as is strain and earthquake relaxation in volcanic areas. The obvious component in most of these endogenetic processes is crustal shortening or tension, and for this reason the apparatus most often employed to simulate effects is the pressure box.

Experimental work on crustal deformation has a long history – for example Avebury[1] in 1903 used layers of carpet and sand beneath plate glass to simulate mountain building, and Howe,[2] some two years earlier, compressed a variety of substances with a screw-driven piston to simulate laccolith formation and associated crustal deformation.

Descriptions of apparatus appear in the main geological journals, and a good bibliography is given in Charlesworth and Passero[3] under the heading of tectonics. A review of experimental work in structural geology is contained in a paper by Currie[4] in 1966, in which theory, apparatus and materials are discussed.

The pressure box is usually quite small – less than a metre in length – and correspondingly narrow and shallow. It is commonly activated by a screw thread, turned by a motor or by hand, which moves a plate against layers of sediment.

Materials used for sediment layers are wide ranging and include modelling clay, field clays, wax, sand, plaster of Paris, coal dust, plaseline, gelatine, glycerine, vaseline, lubricating oil, stucco, rubber sheet, corn syrup and roofing tar. Different properties in layers are produced by varying temperature, wetting or drying, lubricating and applying vertical pressure with an overburden of lead shot or mercury.

Demonstration of one or two common processes may be executed with little sophistication using the materials already in store for other experiments described in this work, but, since they will be mixed together, laboratory preparation of material and clearing up are more time consuming than usual, and it is advisable not to use very clean, well-sorted store material.

The pressure box employed was 60 cm long by 10 cm wide and 25 cm deep. It was strongly constructed of hardwood, reinforced at the edges with angle-aluminium. The front was of safety glass, which was also removable for cleaning purposes. The drive mechanism was a $\frac{1}{4}$ h.p. electric motor running at 1725 rpm, reduced by a belt and pulley system to turn a threaded collar at a rate of 26 rpm. This forced through a heavy-duty car-jack worm at a rate of 11 cm/min to force the pressure plate forward (Figure 11.1). The pressure plate was held firm by a removable guide rod passing through its upper portion.

Laying clay uniformly in sediment sequences is problematical. It cannot be laid dry, and if laid as a wet plastic it disturbs the layers beneath. If the pressure box (which is not water tight) is seated in a water tank as shown in Figure 11.2, sediments can be fed into water and they tend to settle fairly uniformly. Stirring or agitating encourages equal distribution. With sand, the most effective technique has been found to be sprinkling a clean, dry quantity through a sieve

Figure 11.1. Pressure box and drive mechanism.

Figure 11.2. Pressure box seated in water tank.

mesh designed to fit over the top of the box. Best results with clay, however, were to mix a sludge which was then fed through a funnel and tube system to a point below the water surface at the back of the box where any distortion of the material already laid would not show (Figure 11.3).

Smoothing of dry-laid materials is done with scrapers and brushes. Wet sand can be levelled off under water by sharply tapping the pressure box. Clay has to be allowed to settle slowly - for at least 15 minutes to obtain a few millimetres thickness, or overnight if a thicker layer is required.

Experiment 1

Purpose: To produce folding in unconsolidated sediment layers.

Apparatus: Pressure box, fine sand; clay; water supply.

Procedure: The pressure plate was adjusted to the required position; the guide rod was removed. Water was introduced to the pressure box and trough system to a depth of 7.0 cm. A 1.8 cm thick layer of sand was introduced as a basal sealer, then clay (0.8 cm), sand (0.7 cm), clay (0.3 cm) and sand (0.8 cm) were added in sequence, making up a total series of 4.4 cm in a test block 10 cm wide and 43.2 cm long.

After leaving to settle for 15 minutes, the water was completely siphoned out; the glass plate was cleaned and the guide rod replaced. The position of the strata was marked with grease pencil on the glass (Figure 11.4 (*a*) and (*b*)) and the experiment was started. It was halted for photography and measurement at intermediate pressure plate locations of 1.5, 3.5, 5.7, 8.7 and 11.4 cm.

Controls: $\frac{1}{4}$ h.p. applied for 1.68 minutes with a rate of compression of 11.0 cm/min. Materials: fine sand (0.177-0.0625 mm) and silty clay.

Observations and data: Crustal shortening was seen to occur, as follows.

1.5 cm: folding in lower layers; faulting with overthrust in top sand layer (Figure 11.5).

3.5 cm: second fold and fault formation; no

Figure 11.3. Technician feeding clay as sludge into pressure box.

Figure 11.4 (*a*). Experiment 1 before commencement, showing cleaned-up apparatus with guide bar replaced.

Figure 11.4 (*b*). The position of sediments is marked with grease pencil.

visible distortion ahead of front fault (Figure 11.6).

5.7 cm: third fold and fault formation. A dense basal core is developing under the main fold zone. No visible distortion ahead of third fault (Figure 11.7).

8.7 cm: four fold and fault systems well developed, the fifth just starting ahead. Top clay layer clearly shows uniform folds; middle sand layer is bunching up and lower clay is distorted. The bottom sand forms the basal bulk mass. Height of land mass now 8.8 cm (Figure 11.8).

11.4 cm: sixth fold and fault just commencing ahead.

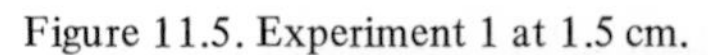

Figure 11.5. Experiment 1 at 1.5 cm.

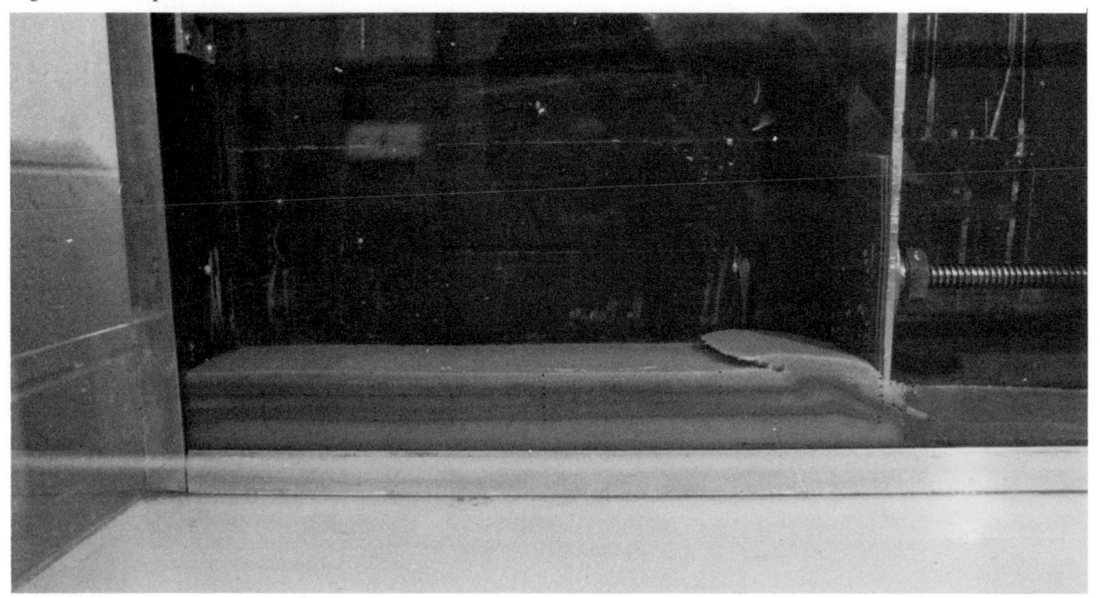

Figure 11.6. Experiment 1 at 3.5 cm.

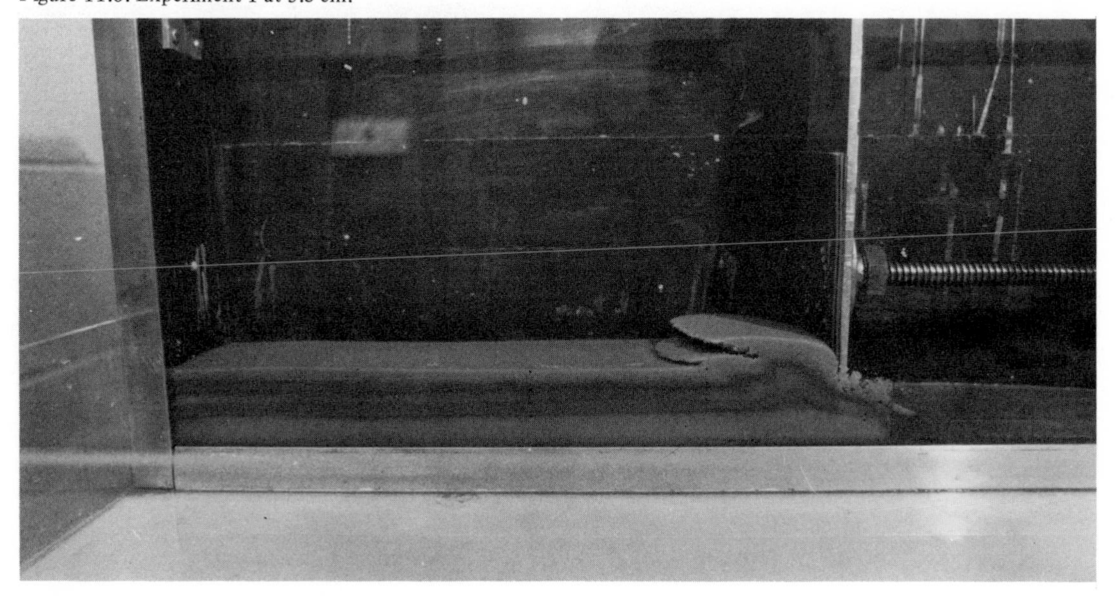

18.5 cm: six full and four rudimentary fold/fault systems exist with a new major fold system imposing upon the bottom clay layer. Still no distortion ahead of leading fault.

The top sand layer is fractured into a set of overthrust faults spaced at 1.7 cm. The top clay layer is folded with two shear zones in the forward folds, and the sands and clay beneath are highly distorted. Crustal shortening is 18.5 cm and increase in elevation is from 4.4 cm to 13.0 cm = 8.6 cm. The visual zone of distortion is 15 cm in length. *Conclusion:* The stratified deposits responded to lateral pressure by movement near the force zone

Figure 11.7. Experiment 1 at 5.7 cm.

Figure 11.8. Experiment 1 at 8.7 cm.

to the right. The initial tendency was evidently to lateral compression accompanied by vertical displacement of a mass a few centimetres wide. The compression-uplift response was essentially a folding process, but while the upper clay stratum took up the fold, the unconsolidated sand above fractured, producing an overthrust fault.

As the crustal shortening continued, the sediment sequence repeated the folding, fracturing process in the forward position, while the rear mass, including the first fold and fault, rose higher. This process recurred for approximately each 1.85 cm of crustal shortening; the material failed with this horizontal distortion and a fold height of about 1.2 cm. The new conformation of the strata was stronger than the *in situ* beds at the foot of the advancing wall, so that the next fold and fault was always in the proximal position, cumulatively forming an imbricated structure. The beds beneath the upper clay layer were compressed without the freedom to fold, except in a rudimentary way, and a root zone of highly distorted sediments resulted.

At the end of the run, a new fold had just begun to the right of the mass indicating a later superimposition of a greater magnitude fold system.

Systems interpretation: The total sediment bed is regarded as a system, with morphological parameters 4.4 cm thick, 10.0 cm wide and 43.2 cm long. The specific weight of submerged fine sand is taken as 0.87 g cm^{-3}, and submerged clay 0.8 g cm^{-3} giving for the series a mass of 2210 g of sand and 380 g of clay.

The energy input to the system is $\frac{1}{4}$ h.p. for 1.68 minutes. Since 1 h.p. = 746 J s^{-1}, $E_k = \frac{1}{4}$ (1.68 x 60 x 746) = 18 799 J. All of the energy is used in doing work within the sediment mat* or is converted, by friction, to heat, and released into the atmosphere. There is no input or output of mass, nor does the system as a whole approach an equilibrium condition; but individual fold/fault subsystems appeared to be stable for part of the experiment, with the zone of optimum shear stress always being at the leading edge of the distorted mass. The system may be canonically represented as shown in Figure 11.9.

Experiment 2

Purpose: To produce a fault system in sand strata.

Apparatus: Pressure box; coarse sand; fine sand; plaster of Paris; base plate.

Procedure: The pressure plate was set at 46.0 cm and the guide rod was removed. A base plate of 3 mm plexiglass, 20 cm long with a bevelled leading edge, was fitted into the box, tightly against the pressure plate. 2160 g of fine sand were saturated in 600 ml of water in which 95 g of plaster of Paris had been mixed, and the excess water was then decanted. The wet sand was laid in the dry pressure box where it was levelled by gentle tamping and smoothed with a scraper to form a 3.0 cm thick layer. It was left for 21 hr, after which 1346 g of coarse, dry sand were laid on top through a spreading sieve to form a 1.9 cm thick layer. The front glass was cleaned and the

* Note that machine efficiency is ignored.

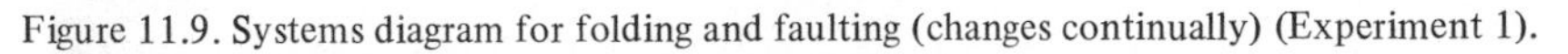
Figure 11.9. Systems diagram for folding and faulting (changes continually) (Experiment 1).

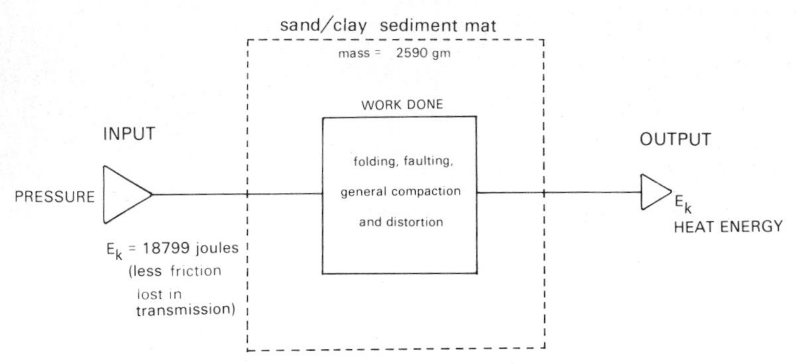

guide rod was re-seated (Figure 11.10). The experiment was halted at 3.6, 4.9, 6.5 and 10.2 cm for measurement and photography.
Controls: The fine sand was made slightly more cohesive and inflexible by addition of plaster of Paris. The pressure plate was moved through 10.2 cm over a period of 0.93 minutes (11.0 cm/min).
Observations and data: The first significant response to pressure was the development of two reverse faults, which joined at the base of the fine sand layer (Figure 11.11). The first fault com-

Figure 11.10. Experiment 2 before commencement. Note base-plate beneath sediments to the right of the model.

Figure 11.11. Experiment 2 soon after commencement. Note effect of base plate which causes failure towards the centre of the model.

menced at the bevelled edge of the base plate, and the right-hand reverse fault then developed, raising a wedge of fine sand into a horst, folding up the coarse surficial sand layer.

3.6 cm: a second left-side thrust fault is developing and both left faults are now undercut by the right reverse fault. Steps are developing at the surface in the coarse sand layer.

4.9 cm: a major reverse fault now occurs from top to bottom of the fine sand layer. Dry coarse sand is simply falling into the fractured fine layer forming a gravity-creep surface – which is

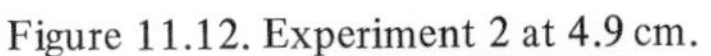

Figure 11.12. Experiment 2 at 4.9 cm.

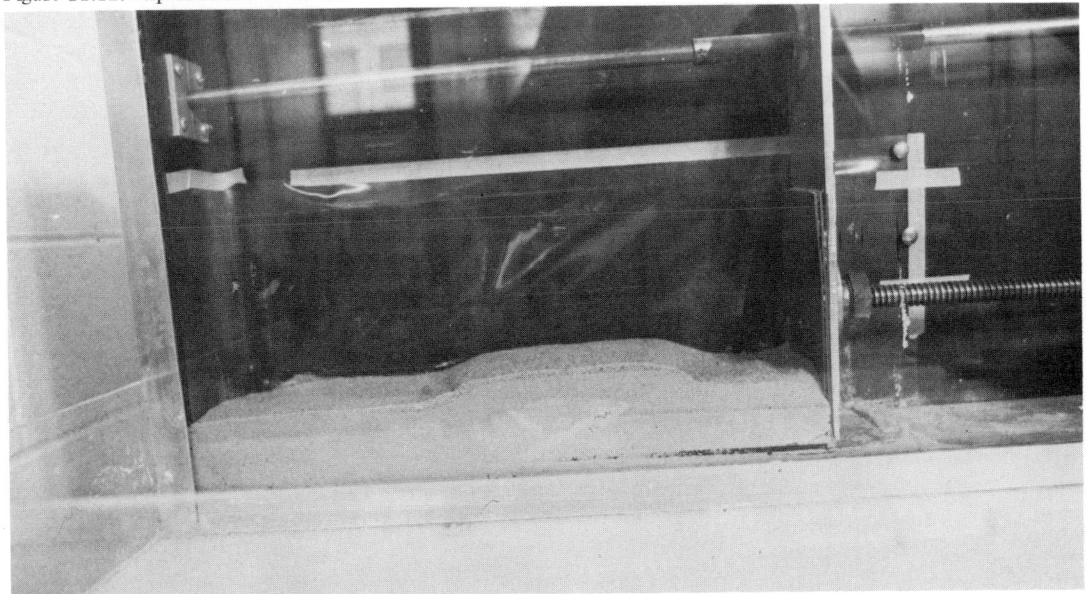

Figure 11.13. Experiment 2 at 6.5 cm. Note the relationship of surface morphology to subsurface conditions.

a mass-wasting process – and effectively smooths out the surface expression of the horst (Figure 11.12).

6.5 cm: the fine sand horst has now risen to about 2.0 cm above the original land surface. Stratification outside the fault zone is still horizontal. A fourth thrust fault has appeared, and the symmetry of faulting is marked at a spacing of one fault per 0.9 cm. The major reverse fault is perfectly planar (Figure 11.13).

10.2 cm (end point): the characteristics of planarity, symmetry and uniformity have persisted to the end point (Figure 11.14). In addition to the major reverse fault, six parallel and equally spaced forward thrust faults occur, each 2–3 cm in length. The main fault is 12 cm long with a 2.5 cm throw, and, with bed shortening of 10.2 cm, the horst-like landform is 22.0 cm in length. Coarse sand has been thrust beneath the upthrow edge of the fine sand layer, with some fracturing and cavitation, and the surface of the model is stepped over the faults.

Conclusion: With a weak but rigid layer of fine sand, lateral pressure will cause failure at a weak zone (here, above the edge of the undersliding base plate) and reverse or thrust faulting will result. The shear properties of the bed, and the pressure applied being generally uniform, a series of small faults of closely equal size and spacing will relax the strain in the bed until the pressure is removed. If the surficial layer is dry, non-cohesive material, it will not shear, but will heap with pressure and creep under gravity to form a stepped surface, reflecting in smoothed form the distortions beneath.

Systems interpretation: As in Experiment 1, there is no input or output of matter to the system, which is clearly defined within the pressure box. Kinetic energy, applied through the pressure plate, is $\frac{1}{4}(0.93 \times 60 \times 746) = 10406.7$ J during the 0.93 minute run, and this is used in compressing and distorting the bed material, some being dissipated as heat.

As in Experiment 1, although the system as a whole is in a state of constant flux while pressure is applied, the small faults, which may be regarded as sub-systems, appear to possess stability, which suggests that the energy of the system is transmitted through the major fault until new threshold conditions are reached, at which time a new minor fault occurs.

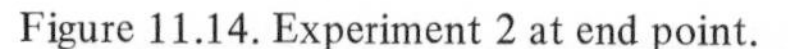
Figure 11.14. Experiment 2 at end point.

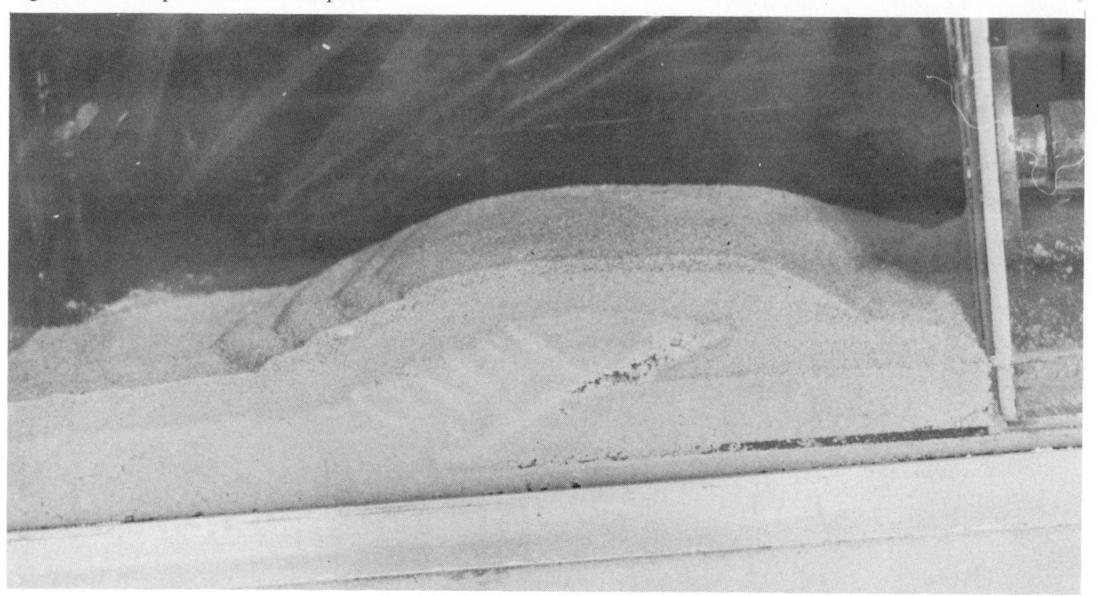

12 Field models

Certain small-scale natural phenomena may be used as models of larger-size landforms; for example, on beaches, rills or gullies of a metre or so in length occur which may be used as models of much larger drainage forms. Occasionally an algae mat of a few centimetres thickness will be left behind as the tide goes out, developing a perfect dendritic drainage net as water seeps from the beach (Figure 12.1). Mud-flats and tidal flats in estuaries frequently develop drainage networks in miniature, and in road cuttings washed-out fines carry such micro-forms as rills, cliffs and pipes. Puddles also frequently develop miniature geomorphic features. These may be taken advantage of as they are discovered, to demonstrate some earthforms and associated processes.

A fairly common situation that offers considerable opportunity for detailed model study is the tailing or sludge mound found near mines and quarries, such as the clays from lead workings at Glendassan, Co. Wicklow (Figure 12.2) and the sludge mounds at Lundrigan's quarry near St John's, Newfoundland.

The St John's mounds have been used for a number of years as a field example of micro-landscape formation, requiring continued re-appraisal since the mounds are constantly being bulldozed away to make room for more waste.

The material is a white clay derived from crushing and washing processes. It is piped as a sludge to settling ponds and periodically shifted to an edge of a pond to facilitate draining (Figure 12.3).[1]

Figure 12.1. Dendrite drainage net formed as tide recedes at Blackrock, Dublin.

Figure 12.2. Clay mound from lead working at Glendassan, Co. Wicklow.

Figure 12.3. Map of settling ponds and sludge mounds at Lundrigan's Quarry, St John's, Newfoundland.

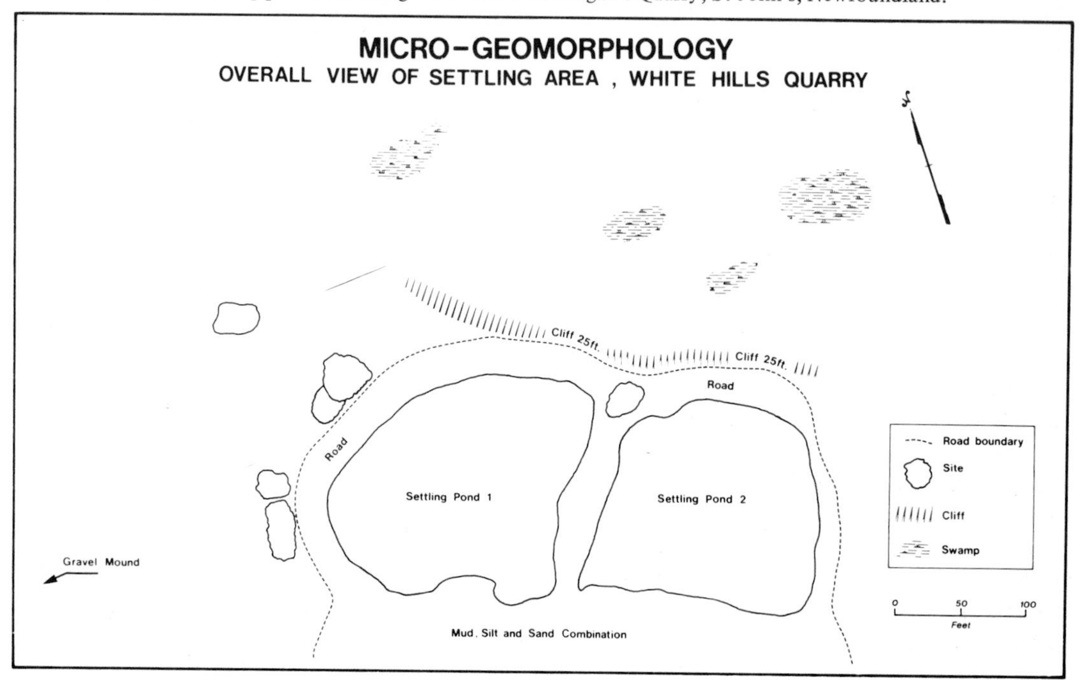

Natural rainfall over a period of time will erode the mounds, producing small-scale features that have much in common with semi-arid landforms. These are frequently so highly dissected that they also display badland features.

The exercises that may be carried out are (*a*) identification of morphological features, (*b*) deduction of processes at work, (*c*) morphometric analysis, (*d*) systems analysis. These require somewhat different approaches and each could be undertaken by a group within a class and brought together afterwards. The remainder of this chapter deals with each of these four points in turn.

12.1 The identification of morphological features

The fieldwork is approached after it has been established that periodic heavy rainfall upon certain types of rock, in the absence of a vegetation cover, will produce semi-arid and badland-type landforms. Semi-arid landforms are characteristic of the south-western areas of the United States, and parts of Mexico. As such landforms are not normally to be seen in wetter areas, their occurrence in analogue form has particular value.

A basic problem immediately arises in identifying the morphological features, involving nomenclature of model forms as distinct from prototype forms. This theme was introduced in Chapter 6 where an approach to the study of semi-arid features was undertaken with the rainfall simulator.

In the present instance, in so far as a clay deposit erodes under natural rainfall, all features may be regarded as natural rather than laboratory induced and controlled. The landform need not therefore be regarded in any sense as a model. If it is studied as a prototype, it will be found that no terms exist to describe many of the micro-forms that occur. If it is used as a model of a larger-scale, semi-arid prototype, terms exist that will describe most if not all significant features, but a conceptual distinction should be made between words that may properly be applied to the small-scale features as well as to their larger counterpart, and words which may only be used because they refer to an analogue of the prototype. This has been done by providing students with a list of landforms upon which they check for *real* or *analogue* usage and *real* (r) or *iconic* (i) usage. (It may be assumed that clay is a prototype bed material. (See Chapter 1.))

The list may include mesa (i), rectilinear slope (r), fan (r), plunge-pool (r), desert hollow (i), alluvial plain (i), vertical-walled water courses (r), debris (r), pediplain (i), dry ravine (i), plateau (i), pediment (i), arroyo (i), parallel slopes (r), residuals (r), upland area (i), butte (i), interfluve (r), gully (i and r), bajada (i), structural terraces (r), divide (r), rounded upland area (i), knife-edged upland area (i), castellated upland area (i), piping (r), playa (i), relict forms (r), rill (r), hoodoo (i), earth pillar (i), amphitheatre valley head (i) and ephemeral stream (r).

Since many of these features are produced, in the micro-form, by a natural process, the most likely criterion to be employed in deciding whether a given element is real or iconic will be that of its size connotation in normal technical usage. This will usually carry much the same connotation as the layman's colloquial or general term from which the technical terms have developed.

The interesting point arises that, if there were no size limitation associated with many of these geomorphic terms, there might be little reason to regard the micro-landforms as models at all. They are not representative of a *specific* larger-sized landform and could represent an example of what Chorley has termed 'unscaled reality'. However, as no formal line has yet been taken on nomenclature in micro-landforms, it seems best to permit size to be a determinant in terminology and to recognize the analogue or iconic concept.

In addition to rationalizing the landforms and their names, students should sketch each feature. This focuses the mind on essentials and fixes the image. It may be necessary to permit the student to refer to sketches in such works as Lobeck.[2] A definition should also accompany the sketch (Figure 12.4).

12.2 Deduction of process

Usually, the field trip will be on a fine day, and the processes of land-sculpting will not be observed. Even if it is raining and the opportunity is taken of observing runoff at work, many aspects of the dynamics will have to be deduced - although use of the rainfall simulator can aid in gathering some empirical information (see Chapter 6).

The early stage of runoff and erosion may be deduced by using the Horton model for overland flow as a guide. At the commencement of rain, infiltration will cause the moisture content of the bed mass to increase to saturation downslope, leaving an unsaturated upper zone free from sheetflow and erosion (Figure 12.5). Rills develop in the lower areas, eroding back into the landmass by headward extension, and bifurcating as the runoff area increases (Figure 12.6). Some rills grow to gullies (Figure 12.7) which may be the equivalent of a prototype arroyo.

Drainage basins usually develop where concave surfaces existed in the original surface, and dendritic drainage patterns evolve. Steep rill or gully sides develop by a process of basal seepage, sapping and slumping and these constant slopes are common in the earlier stages of the developing landform (Figure 12.8).

The dendritic rill system exists on what is probably equivalent to the free-face,[3] which also remains at a constant angle as it retreats. As the landform is reduced, by extending rill systems and free-face retreat, higher areas stand out as plateaus (Figure 12.9), and, at a more advanced stage, isolated heights remain which are similar to buttes or mesas (Figure 12.10). Sometimes piping develops strongly as the rills loose water by rapid percolation, or as water tributary to a rill flows underground just prior to reaching the rill, and collapse features may result (Figure 12.11).

As the free-face, with its dense rill system, resembling badland topography, retreats across the landmass, it leaves a gently rising pediment upon which is spread the mud and debris from rills and gullies. The fans coalesce to form a bajada (Figure 12.12). Larger rills or gullies open out at the mouth to such an extent that their floors become indistinguishable from the pediment, and free-face features develop on the rill walls. The bajada carries surface wash that may run into a lake, which in dry spells has no outlet and resembles a playa, but more generally a stream drains from the lake (Figure 12.12). Scattered across the bajada are larger pieces of debris (boulders) which lie in a specific zone (boulder apron, Figure 12.13).

Occasionally the existence of bedding, with differentiated resistance to erosion, allows the

Figure 12.4. Reproduction from a student's notes in which micro-landforms are related to large-scale reality.

landform	prototype / iconic or analogue	definition	graphics
divide	P	a mountain range or area of high land or rise of land separating one drainage system from another	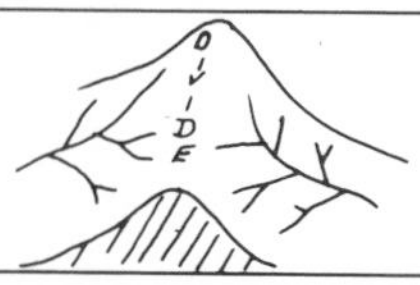
mesa	A	table-topped hill or mountain bordered on all sides by cliffs representing the remnant of a formerly extensive layer of resistant rock.	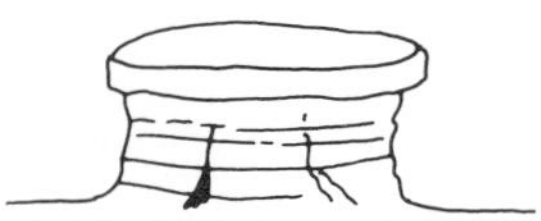
rectilinear slope	P	a slope bounded by straight sides.	

Figure 12.5. A zone of no-erosion in a micro-landform situation, with a rill in lower land (Lundrigan's Quarry).

formation of contorted forms, and of caves and hoodoos. Eventually most of the landforms will disappear, leaving a bland, featureless plain - a panplain - with a few residuals (Figure 12.14).

12.3 Morphometric analysis

Parts of the landmass develop into drainage basins which, in the earlier stages, contain a well-formed drainage net, lending themselves well to morphometric analysis. They offer an opportunity to use a variety of data-collecting techniques and to apply principles of quantitative analysis quickly and conveniently.

Measurement of the selected basin requires the use of a level and stadia-rod, or some substitute such as a spirit level or Brunton compass. Metre-sticks, stakes, rope and a nylon surveyors' tape are also required (nylon is less likely than metal to damage micro-features).

The basin is staked out and ropes or thick cords are raised above it. They are marked off in metres and decimetres and are horizontal and at

Figure 12.6. Rills eroding back into a highland area (Lundrigan's Quarry).

Figure 12.7. Gully formed by runoff and erosion (Lundrigan's Quarry).

Figure 12.8. Parallel slope angles in rill and gully development (Lundrigan's Quarry).

Figure 12.9. Plateau and free-face (Lundrigan's Quarry).

Figure 12.10. Isolated heights in area of advanced erosion (Lundrigan's Quarry).

Figure 12.11. Water erodes in subsurface lines, and collapse features occur (Lundrigan's Quarry).

right-angles to each other (Figure 12.15). Students then take readings of offsets with metre-sticks along the two axes, recording the distances to basin perimeter, and depths to perimeter and significant parts of the basin (Figure 12.16). Data sheets can be provided, but at least one person should mark values on a sketch map to facilitate reconstruction of the form.

The position of the main channel can be located by this technique and other members of the net can be sketched in by eye - a procedure requiring good judgement, and although students should be given the task of field sketching, the instructor should make a sketch for reference. The basin may be tied in with surrounding features, and horizontality of lines checked with the level.

Two representations of the basin will be obtained - a preliminary field sketch and a later topographic map (Figure 12.16). The drainage net will not have the accuracy on the final map that would be obtained from comprehensive measurement, but topologically it should be accurate, and segment lengths should not

Figure 12.12. Fans coalesce to form a bajada (Lundrigan's Quarry).

Figure 12.13. Boulder apron on bajada (Lundrigan's Quarry).

Figure 12.14. A bland featureless plain, with a few residuals (Lundrigan's Quarry).

Figure 12.15. Drainage basin staked out for measurement, with horizontal lines at right angles to each other.

Figure 12.16. Topographic map of micro-drainage basin. Contour interval 10 cm.

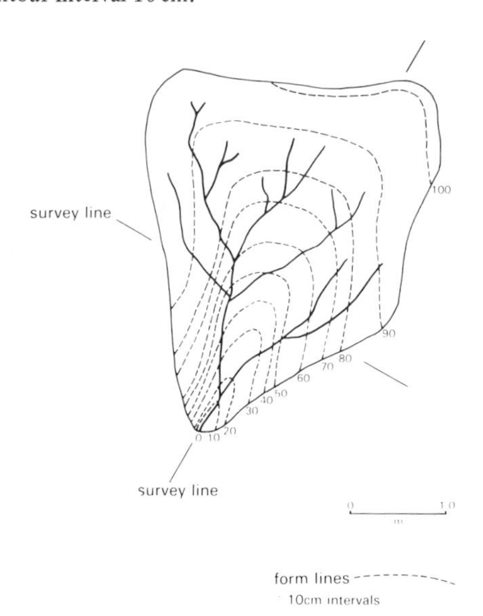

incorporate much error. If necessary, photographs can be taken from a step-ladder.

Example (primary and derived data)
Basin length (L_B) 4.09 m (field measurement)
Basin width (W_B) 3.00 m (field measurement)
Basin area (A) 7.5 m^2 (polar planimetry from final map)
Basin perimeter (P) 10.8 m (opisometry from final map)
Circularity ratio (C) 0.81 (from map)
Basin relief (H) 1.03 m (field data)
Mean basin relief ($\bar{H}$) 0.85 m (from map)
Basin order (Strahler) 3 (map network)
Bifurcation ratio R_B (weighted mean) 3.08 (map network)
Drainage density (D) 1.7 (from map)
Main stream channel near mouth
width (w) 0.12 m (field measurement)
depth (d) 0.06 m (field measurement)
width/depth ratio (w/d) 0.2 (field measurement)

Segments of order u (N_u)

Order u	N_u
1	12
2	5
3	1

Mean length of segments of order u (L_u)

u	L_u	ΣL_u (cm)
1	52.2	630
2	78.0	390
3	230.0	230
		ΣL 1250 cm

This example of morphometric analysis uses a relatively uncomplicated drainage basin which incorporates several of the most common morphometric concepts. Contemporary basins on site may be compared and basins from previous trips may be re-introduced. The validity of Horton's network laws should be checked in all cases.

The principles of morphometric analysis may be extended to some of the other features. For instance, measurements may be taken of the overall extent of the major components - gullies, free-face, bajada, plateau - or detailed examination may be made of badland features, or playas.

12.4 Systems analysis

The gross dynamics of parts of the landform are best studied within a systems framework. If a drainage basin is chosen, as seems most practical, the morphological subsystem will already be to hand from the morphometric analysis, and it will remain only to assess the mass and energy components of the process-response system (Figure 12.17).

With uniform bed material and a lack of vegetation, the system is relatively simple, but in this field study, in contrast to the laboratory study of Chapter 6, accurate rainfall data will not usually be available.

It is uncommon for the workers at the site to know exactly when a particular mound was dumped or modified, and approximate time and quantity values have to be used. Also, the size of raindrops and cloud height will not normally be known, so that energy calculations are approximate. As the exercise is to illustrate techniques and concepts rather than to describe, the use of estimated values seems to be justified.

Mass input (water): with a basin area of 7.5 m^2 (75 000 cm^2) and a precipitation depth for $5\frac{1}{2}$ months of 53.5 cm* the total mass water input will be 4 012 500 cm^3 (4 012 500 g).

Mass input (sediment): if it can be assumed that the basin landform was originally a prism, with section defined by the basin perimeter, its original volume would have been $A \times H = 75\,000 \times 103 = 7\,725\,000$ cm^3. Since the volume of the prism remaining is $A \times \bar{H} = 75\,000 \times 85 = 6\,375\,000$ cm^3, the eroded portion would be $7\,725\,000 - 6\,375\,000 = 1\,350\,000$ cm^3. With a bed material density of 2.53 g cm^{-3} this represented 3 415 500 g.

* Note that evaporation may be ignored for runoff calculations owing to the brief time the water from a given storm remains in the catchment.

This is the type of reasoning used in determination of the hypsometric integral of drainage basins and for assessing their stage of development. There is little possibility that the initial landform in the present basin was a perfect prism - it was almost certainly a random declivity on the face of a mound. However, it can be said that sediment input is less than 3.416×10^6 g and this establishes an order of magnitude.

Energy input (water):
impact energy* $(E_k) = \frac{1}{2}mv^2 = \frac{1}{2}(4\,012\,500 \times 650^2) = 8.48 \times 10^{11}$ ergs $= 8.48 \times 10^4$ J;
runoff energy $(E_p) = m \times \bar{H}/10\,197.16 = 4\,012\,500 \times 85/10\,197.16 = 3.34 \times 10^4$ J.

Energy input (sediment):
$E_p = m \times \bar{H}/10\,197.16$ J, which will be less than $3\,415\,500 \times 85/10\,197.16 = 28\,470$ J or $<2.84 \times 10^4$ J.

As the water throughput acts within the morphological subsystem, it dislodges quantities of bed material uninhibited by a vegetation cover. The quantity of solids moving through the system is probably much less than the 3.416×10^6 g derived from an eroding prism - very probably half this value would be more appropriate, and, consequently, the potential energy of the sediment will be half the 2.84×10^4 J calculated. The relatively high value of raindrop impact energy is notable.

* Note that this represents only the gross impact energy. To relate this to runoff and erosion see Wischmeier and Smith.[4] The velocity of 6.5 m s^{-1} has been selected as representative of local, characteristically small, raindrop diameter.

Many subsystems are identifiable in the basin system, and, if the main channel has opened out, the major components of the landform may exist. For instance, there may be pediment present for which factors to be taken into account will be slope, rate of flow and thickness of sheetwash, sediment load, infiltration rate and shear strength of the bed material.

If a high-density rill erosion face exists (the free-face) the regulators and process controls will be related to the rills - or channelled flow - with their interfluves. Again there will be a characteristic slope, which will relate to rill discharge, bed shear strength, sediment load and so on. Whether this subsystem is dependent upon a water table within the landform, or upon saturated surface layers cannot be induced on *a posteriori* grounds. If it is water-table related, a set of linkages similar to those portrayed in Figure 6.18 may exist. If it is primarily a saturated surface-layer control, as has on occasion been determined in the field, the relationships between the channel and its feed-water will be different.

In this type of field study it should be borne in mind that it is not a field *experiment*, in that the observers have not set up a controlled situation in which variables are few and are known. Consequently, the interpretation of process in detail is as open to debate as is slope development in prototype situations. It is not advisable, therefore, to suggest linkages in any but the most speculative way, nor without reference to the literature of controversy. It follows that the natural field model,

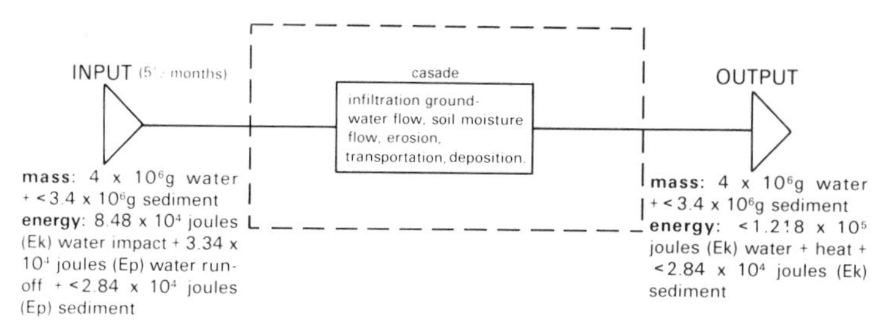

Figure 12.17. Canonical diagram for micro-drainage basin. (The morphological subsystem will include rills, X_c zone and gullies.)

which is one of the most dramatic demonstration aids, has to be approached with particular care.

Some aspects of processes in the total landform can be treated with confidence. The zone of no erosion, where infiltration exceeds runoff, is essentially a feeder zone for the rill system of the free-face. As the free-face extends within a basin or on a hill, the zone of no erosion diminishes, and, as a consequence, a state of equilibrium could be expected between the three components (Figure 12.18).

It is also observed that the rill system will eventually erode the entire landmass, and, consequently, the negative feedback of Figure 12.18 does not completely explain the situation. Precipitation falls also upon the free-face, and the rills will continue to erode by virtue of free-face water input, encroaching continually upon the zone of no erosion.

As the free-face recedes, the pediplain extends and represents a high-entropy surface - the condition of equifinality that all landforms tend towards. Energy will be randomly distributed as water flows irregularly over the surface, randomly depositing sediment. This energy is almost entirely kinetic and is very reduced in quantity from that of the initial stages.

Figure 12.18. Negative-feedback loop relating infiltration in X_c zone, seepage and surface flow, and headward extension of rills.

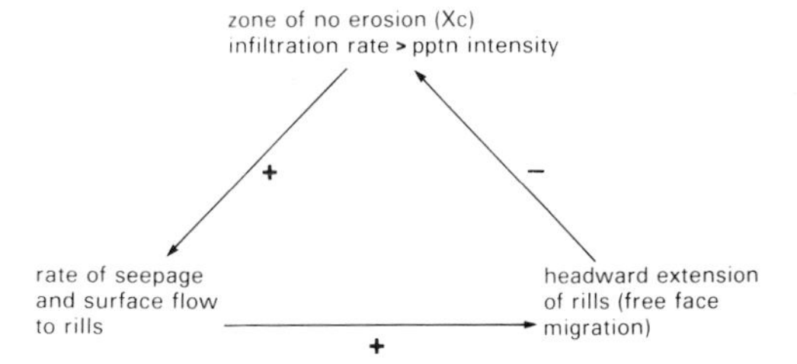

Appendix

The studies in this book use a number of granular materials. In Chapter 5 an organic field soil is used. It should be fairly cohesive and contain a good range of particle sizes.

A 2.0-5.0 mm angular, granitic gravel was used in the spring-line experiment of Chapter 8.

Coarse, medium 'A' and fine sands of sizes 1.0-0.5; 0.5-0.177; 0.177-0.0625 mm respectively were used in some flume experiments. The medium 'A' was the material used in the wind tunnel experiments of Chapter 10. It was very well sorted and washed. The fine sand was also used in the pressure box experiments of Chapter 11.

The most extensively used sand was a 'medium' sand of size 0.074 to 0.5 mm (not to be confused with the medium sand 'A' which was cleaner, and lacked the finer grades of the ordinary medium sand, Figure A.1).

The *clay* of the pressure box experiments in Chapter 11 was a sandy silt with a small percentage of clay.

Figure A.1.

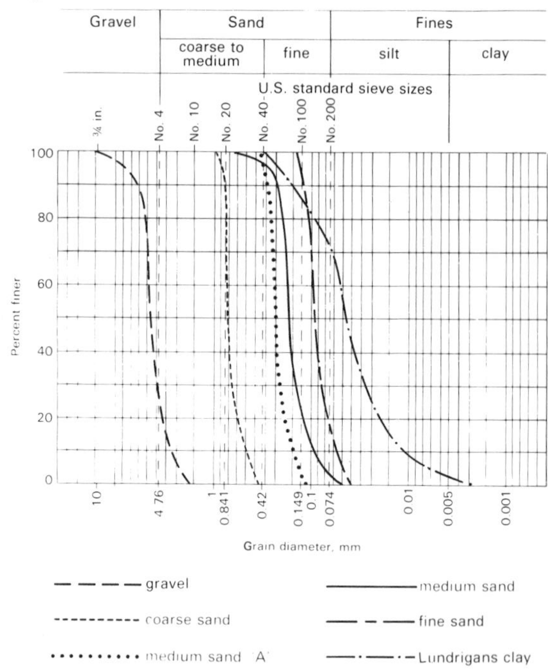

References

Chapter 1

1. Ambler, R., Adams, J. S. & Gould, P. (1971). *Spatial organization* (p. 45). Prentice-Hall, N.J.: 587 pp.
2. Balchin, W. G. V. & Richards, A. W. (1952). *Practical and experimental geography* (see Introduction). Methuen, London: 136 pp.
3. H.M.S.O. (1972). *New thinking in school geography* (p. 94). Department of Education and Science, Education Pamphlet 59. H.M.S.O., London: 99 pp.
4. Gopsill, G. H. & Beesley, F. (1965). *Practical geography: exercises with models and improvised equipment* (see Introduction). Macmillan, London: 96 pp.
5. Morgan, M. A. (1967). Hardware models in geography (p. 727). In Chorley & Haggett (eds.), *Models in geography*. Methuen, London: 816 pp.
6. Harvey, D. (1969). *Explanation in geography* (p. 145). Arnold, London: 521 pp.
7. Haggett, P. & Chorley, R. J. (1967). Models, paradigms and the new geography (p. 21). In Chorley & Haggett (eds.), *Models in geography*. Methuen, London: 816 pp.
8. Haggett, P. & Chorley, R. J. (1967). (p. 25) ibid.
9. Chorley, R. J. (1964). Geography and analogue theory (p. 129). *Annals of the Association of American Geographers*, **54**, 127–137.
10. Chorley, R. J. (1964). (p. 136) ibid.

Chapter 2

1. Chorley, R. J. & Kennedy, B. A. (1971). *Physical geography – a systems approach* (see especially Preface and pp. 1–10). Prentice-Hall, London: 370 pp.
2. Dury, G. H. (1981). *An introduction to environmental systems.* Heineman, London: 366 pp.
3. Chorley, R. J. (1967). Models in geography (p. 63). In Chorley & Haggett (eds.), *Models in geography*. Methuen, London: 816 pp.
4. Harrison, R. (1969). Communication theory (p. 64). In Wiman & Meierhenry (eds.), *Educational media: theory into practice.* Merrill, Ohio: 283 pp.

Chapter 3

1. Giles, R. V. (1956). *Theory and problems of fluid mechanics and hydraulics* (p. 50 *et seq.*). Schaum College Outline Series, N.Y.: 274 pp.
2. Vallentine, H. R. (1967). *Water in the service of man* (p. 162 *et seq.*). Pelican, Harmondsworth: 224 pp.
3. Kellerhals, R., Neill, C. R. & Bray, D. I. (1972). *Hydraulic and geomorphic characteristics of rivers in Alberta.* River Engineering and Surface Hydrology, Report 72-1, Research Council of Alberta: 52 pp.
4. Henderson, F. M. (1966). *Open channel flow* (p. 489). Macmillan, N.Y.: 522 pp.

Chapter 4

1. Slivitzky, M. (1976). A perspective (p. 4). In *Hydrological Events.* National Research Council Canada. Associate Committee on Hydrology. Winter 1976.

   Hjulstrom, F. (1935). Studies of the morphological activity of rivers as illustrated by the river Fyris. *Bulletin of the Geological Institute of Uppsala*, **25**, 221–527.
3. Lacey, J. A. (1970). *Humber tidal model.* British Transport Docks Board (pamphlet) and correspondence.
4. Bagnold, R. A. (1947). Sand movement by waves: some small-scale experiments with sand of very low density. *Journal of the Institute of Civil Engineers*, Paper 5554, 447–469.
5. Yoxall, W. H. (1969). The relationship between falling base level and lateral erosion in experimental streams. *Bulletin of the Geological Society of America*, **80**, 1379–1384.
6. Allen, J. (1947). *Scale models in hydraulic engineering* (p. 380). Longmans Green, London: 407 pp.
7. Allen, J. (1947). (p. 381) *ibid.*

8. Laws, J. O. & Parsons, D. A. (1943). The relation of raindrop size to intensity. *Transactions of the American Geophysical Union*, **24**, 452–460.

Chapter 5

1. Pearson, J. E. & Martin, G. E. (1957). *An evaluation of raindrop sizing and counting techniques.* Illinois State Water Survey. University of Illinois Science Report 1: 116 pp.
2. Hayward, J. A. (1967). Plots for evaluating the catchment characteristics affecting soil loss. *Journal of Hydrology (N.Z.)*, **6**, 2, 120–137.
3. Mutchler, G. K. & Herschmeier, L. F. (1965). A review of rainfall simulators. *Transactions of the American Society of Agricultural Engineers*, **8**, 4, 67–68.
4. Chery, D. L. (1963). Runoff studies with a physical model of a watershed. *Transactions of the American Geophysical Union*, **44**, 869–870.
Chery, D. L. (1966). Design and tests of a physical watershed model. *Journal of Hydrology*, **4**, 224–235.
5. Mutchler, C. K. (1965). Water drop formation from capillary tubes. *U.S.D.A. Agricultural Research Service*, 41-107.
6. Chow, V. T. (1967). The progress of hydrology. From *Isotopes in hydrology.* International Atomic Energy Agency, Vienna, 3–20.
Chow, V. T., (1968). R and D of a watershed experimentation system. *Jets Journal*: 3 pp.
Chow, V. T. (1969). Room with a built-in deluge. *Life*, **66**, 22, 77–78.
7. Bertrand, A. R. & Parr, J. F. (1960). *Development of a portable sprinkling infiltrometer.* 7th International Congress of Soil Science. Madison, Wisconsin.
8. Meyer, L. D. (1960). Use of a rainulator for runoff plot research. *Proceedings of the Soil Science Society of America*, **24**, 319–322.
9. Anderson, J. U., Derr, P. S. & Bailey, O. F. (1969). *The use of a rainfall simulator to study soil-water relationships in semi-arid rangeland.* New Mexico Agricultural Experimental Station, Las Cruces, New Mexico.
10. Würm, A. (1936). Morphologische Analyse und Experiment; I und II. *Zeitschrift fur Geomorphologie*, **9**, 1–24/57–87.
11. Hudson, N. (1971). *Soil conservation.* Batsford, London: 320 pp.
12. Holland, D. J. (1964). *Rain intensity frequency relationships in Britain.* Meteorological Office Hydrology Memo 33: 28 pp.
13. Anderson, J. U., Derr, P. S. & Bailey, O. F. (1969). *op. cit.*
14. Yen, B. C. & Chow, V. T. (1969). A laboratory study of surface runoff due to moving rainstorms. *Water Resources Research*, **5**, 5, 989–1006.
15. Bisal, F. (1960). The effect of raindrop size and impact velocity on sand splash. *Canadian Journal of Soil Science*, **40**, 242–245.
16. Laws, J. O. (1941). Measurement of the fall velocity of water drops and raindrops. *Transactions of the American Geophysical Union*, **22**, 709–721.
17. Chow, V. T. & Harbough, T. E. (1966). Raindrop production for laboratory watershed experimentation. *Journal of Geophysical Research*, **70**, 6111–6119.
18. Robertson, A. F., Turner, A. K., Crow, F. R. & Ree, W. O. (1966). Runoff from impervious surfaces under conditions of simulator rainfall. *Transactions of the American Society of Agricultural Engineers*, **9**, 3, 343–346 and 351.
19. Ellison, W. D. (1945). Some effects of raindrops and surface flow on soil erosion and infiltration. *Transactions of the American Geophysical Union*, **26**, 3, 415–429.
20. Bisal, F. (1960). *op. cit.*
21. Laws, J. O. & Parsons, D. A. (1943). The relation of raindrop size to intensity. *Transactions of the American Geophysical Union*, **24**, 452–460.
22. Pearson, J. E. & Martin, G. E. (1957). *op. cit.*
23. Bryan, R. B. (1969). The relative erodibility of soils developed in the Peak District of Derbyshire. *Geografiska Annaler,* **51**, 145–159.

Chapter 6

1. Horton, R. E. (1945). Erosional development of streams and their drainage basins; hydrological approach to quantitative morphology. *Bulletin of the Geological Society of America*, **56**, 275–370.
2. Strahler, A. N. (1952). Hypsometric (area-altitude) analysis of erosional topography. *Bulletin of the Geological Society of America*, **63**, 1117–1142.

Chapter 7

1. Anderson, E. W. (1969). *Hardware models in geography teaching.* Teaching geography, No. 7. The Geographical Association, Sheffield: 11 pp.
2. Schwarz, M. L. (1968). Stream table construction and operation (pp. 1066–1070). In

Fairbridge (ed.), *Encyclopedia of Geomorphology*. Reinhold, N.Y.: 1295 pp.
3. Beckway, G. (1967). *Stream table investigations.* Laboratory manual for the earth-science stream table. Hubbard Scientific Company, Illinois: 31 pp.
4. Györke, O. (1973). *European hydraulic laboratories: a survey*. UNESCO, Paris: 128 pp.

Chapter 8

1. Blench, T. (1969). *Mobile-bed fluviology*. University of Alberta Press, Edmonton: 168 pp + appendices.
2. Leopold, L. B., Wolman, M. G. & Miller, J. P. (1964). *Fluvial processes in geomorphology*. Freeman, San Francisco: 522 pp.
3. Gregory, K. J. & Walling, D. E. (1973). *Drainage basin form and process: a geomorphological approach.* Arnold, London: 456 pp.
4. Schumm, S. A. (1977). *The fluvial system.* Wiley, N.Y.: 338 pp.
5. Melhorn, W. N. & Flemal, R. C. (1975). *Theories of landform development.* State University of New York, N.Y.: 306 pp.
6. Morisawa, M. (1968). *Streams, their dynamics and morphology*. McGraw-Hill, N.Y.: 175 pp.
7. Chorley, R. J. & Kennedy, B. A. (1971). *Physical geography – a systems approach.* Prentice-Hall, London: 370 pp.
8. Rayner, J. N. (1972). *Conservation, equilibrium and feedback applied to atmospheric and fluvial processes.* Commission on College Geography Resource Paper No. 15. Association of American Geographers, Washington: 23 pp.

Chapter 9

1. Balchin, W. G. V. & Richards, A. W. (1952). *Practical and experimental geography.* Methuen, London: 136 pp.
2. Schwarz, M. L. (1968). Stream table construction and operation (pp. 1066–1070). In Fairbridge (ed.), *Encyclopedia of Geomorphology*. Reinhold, N.Y.: 1295 pp.
3. King, C. A. M. (1966). *Techniques in geomorphology*. Arnold, London: 342 pp.
4. King, C. A. M. (1972). *Beaches and Coasts*, 2nd edition. Arnold, London: 570 pp.
5. Bascom, W. (1964). *Waves and beaches. The dynamics of the ocean surface*. Anchor Books, Doubleday, N.Y.: 267 pp.
6. Vallentine, H. R. (1967). *Water in the service of man.* Pelican, Harmondsworth: 224 pp.
7. Ritter, D. F. (1978). *Process geomorphology.* Brown, Iowa: 603 pp.
8. Stong, C. L. (1968). How to make a wave machine to simulate the building and destruction of beaches. *Scientific American*, **219**, 6.
9. Morgan, M. A. (1967). Hardware models in geography. In Chorley & Haggett (eds.), *Models in Geography*. Methuen, London: 816 pp.
10. Chorley, R. J. & Kennedy, B. A. (1971). *Physical geography – a systems approach.* Prentice-Hall, London: 370 pp.

Chapter 10

1. Bagnold, R. A. (1941). *The physics of blown sand and desert dunes.* Methuen, London: 265 pp.
2. King, C. A. M. (1966). *Techniques in geomorphology*. Arnold, London: 342 pp.
3. Cooke, R. U. & Warren, A. (1973). *Geomorphology in deserts.* Batsford, London.
4. Ritter, D. F. (1978). *Process geomorphology.* Brown, Iowa: 603 pp.

Chapter 11

1. Avebury, Lord (1903). An experiment in mountain building. *Quarterly Journal of the Geological Society of London*, **59**, 235, 3, 348–355.
2. Howe, E. (1901). *Experiments illustrating intrusion and erosion.* U.S. Geological Survey 21st Annual Report, Part 3: 291–303.
3. Charlesworth, L. J. & Passero, R. N. (1973). *Physical modelling in the geological sciences: an annotated bibliography.* Council on Education in the Geological Sciences Programs Publication no. 16, American Geological Institute: 84 pp.
4. Currie, J. B. (1966). Experimental structural geology. *Earth-Science Reviews*, **1**, 1, 51–67.

Chapter 12

1. Yoxall, W. H. (1976). The use of microfeatures in landform study. *Interaction: Memorial University of Newfoundland,* **2**, 2, 2–4.
2. Lobeck, A. K. (1939). *Geomorphology.* McGraw-Hill, N.Y.: 731 pp.
3. Wilson, L. (1968). Slopes (p. 1007 – after L. C. King). In Fairbridge (ed.), *Encyclopedia of Geomorphology*. Reinhold, N.Y.: 1295 pp.
4. Wischmeier, W. H. & Smith, D. D. (1958). Rainfall energy and its relationship to soil loss. *Transactions of the American Geophysical Union*, **39**, 285–291.

General sources of particular merit

The Journal of Geological Education (National Association of Geology Teachers, Inc.) which has published many techniques for dynamic model use.

Charlesworth, L. J. & Passero, R. N. (1973). *Physical modelling in the Geological Sciences: an annotated bibliography*. Council on Education in the Geological Sciences Programs Publication no. 16, American Geological Institute: 84 pp.

Heller, R. L. (ed.) (1962). *Geology and earth-science source-book for elementary and secondary schools*. Holt, Rinehart and Winston, N.Y.: 496 pp.

Hubbert, M. K. (1937). Theory of scale models as applied to geologic structures. *Bulletin of the Geological Society of America*, **48**, 1459–1520.

Whalley, W. B. (1976). *Properties of materials and geomorphological explanation*. Oxford University Press: 60 pp.

Index